Mathematics in Mind

The monographs and occasional textbooks published in this series tap directly into the kinds of themes, research findings, and general professional activities of the **Fields Cognitive Science Network**, which brings together mathematicians, philosophers, and cognitive scientists to explore the question of the nature of mathematics and how it is learned from various interdisciplinary angles. Themes and concepts to be explored include connections between mathematical modeling and artificial intelligence research, the historical context of any topic involving the emergence of mathematical thinking, interrelationships between mathematical discovery and cultural processes, and the connection between math cognition and symbolism, annotation, and other semiotic processes. All works are peer-reviewed to meet the highest standards of scientific literature.

More information about this series at http.s://link.springer.com/bookseries/15543

Robert K. Logan • Izabella Pruska-Oldenhof

A Topology of Mind

Spiral Thought Patterns, the Hyperlinking of Text, Ideas and More

Springer

Robert K. Logan
Department of Physics
University of Toronto
Toronto, ON, Canada

Izabella Pruska-Oldenhof
School of Image Arts
Ryerson University
Toronto, ON, Canada

ISSN 2522-5405 ISSN 2522-5413 (electronic)
Mathematics in Mind
ISBN 978-3-030-96435-1 ISBN 978-3-030-96436-8 (eBook)
https://doi.org/10.1007/978-3-030-96436-8

This Springer imprint is published by the registered company Springer Nature Switzerland AG
The registered company address is: Gewerbestrasse 11, 6330 Cham, Switzerland

Preface

What This Book Is About

This book covers many diverse topics related in varying degrees to mathematics in mind. We will examine mathematics in mind from the perspective of the spiral, cyclic, and hyperlinked structures of the human mind in terms of its language, its thoughts, and its various modes of communication.

We report on two studies we have made of the mathematical and topological structures of thought and communication that the human mind creates: one of the spiral and cyclic structures found in human thought, the arts, and social interactions and the other of hyperlinking of separate bodies of information. This book is divided into four parts.

In Part I, we discuss what mathematics in the mind entails and use abductive reasoning to suggest that the human mind, verbal language, and mathematics in mind (in the form of set theory) are interconnected and emerged simultaneously as pre-human hominids made the transition to become fully human, Homo sapiens. We show how the mind capable of conceptualization, verbal language, and mathematical thinking forms a supervenient emergent dynamic system making use of complexity theory. We explore the cognitive origins of mathematical thinking of the human mind and its relation to the emergence of spoken language. We will also examine the emergence of mathematical notation and its impact on mathematics, education, science, accounting, and other applications of mathematics.

We show how spoken language and the ability to conceptualize and thus to plan and to think about things that are not available in the immediate here and now arose through the linking of concepts with percepts and the ability of humans to think in terms of set.

We also show how hyperlinking between spoken words and visual signs led to the origin of the written word and, as we show, arose out of accounting. We examine the importance of mathematical notation for advancing scientific and mathematical thinking while at the same time make a careful distinction between mathematics and science as well as logic and science.

In Part II, we examine the spiral and cyclic topological structures of human thought. The spiral structure is a ubiquitous and universal form that one finds in nature in physical structures of whirlpools, tornados, hurricanes, and galaxies throughout the universe. Spirals are also a ubiquitous feature of biological life from the molecular forms such as the double helix of DNA in the genetic makeup of all living things to the patterns of biological forms like the phyllotactic spirals in the arrangement of seeds in a sunflower and of the petals of flowers, the branches of trees, and the shells of mollusks. Spiral structures are also found in various domains of human study and cultural expression including media, the evolution of technology, history, and the thoughts of many scholars such as Marshall McLuhan, Sigmund Freud, and I.A. Richards; certain poets and writers such as Edgar Allan Poe, T.S. Eliot, James Joyce, Gertrude Stein, and Ezra Pound; certain artists such as Wyndham Lewis and the Vorticism Art Movement. We also study the impact of cyclic structures on human thought patterns which are closely related to the spiral. The spiral executes cyclic or rotational motion around a reference point with the radius of rotation either increasing or decreasing. Wave motion is an example of cyclic motion in the physical world. We will examine examples of cyclic structures in the thought patterns of Giambattista Vico's corso and ricorso, the dialectics of G.W.F. and Karl Marx, the eternal return of Friedrich Nietzsche and Mircea Eliade and the notion of the recurring universe of certain cosmologists. As part of our study, we also examine the spiral and cyclic structures in both the abiotic physical world and in the biosphere as a way of demonstrating the universality of these topological structures.

In Part III, we study the role of hyperlinking in human communication, thought, and scholarship. Hyperlinking is basically the linkage of one body of information with another, which came into prominence with the World Wide Web because of the ease with which one body of information can be linked to another. The roots of hyperlinking and hypertexting found in digital media can be traced back to the origin of human communication and its subsequent expression in terms of the origin of the spoken word and its use particularly in epic poetry.

Our choice of the term "hyperlink," which is common in digital media vernacular, is intentional. Firstly, the etymological root of "hyper" is in the Greek word *huper*, which means "over, beyond." Because our interest in this section of our book is the thought process itself, we want to distinguish the linking or connecting processes in the physical world from linking that takes place in mental processes of human thinking, remembering, dreaming, imagining, etc. We are fully aware that the human brain and its dynamic processes are inseparable from the physical world. Secondly, from perspective of neurophysiology, thought processes are electric, electrochemical to be more specific, and rely on electrochemical activation of neurons, which fire neurotransmitters that activate the electrochemical process in other neurons in a domino-like fashion across neuropathways. Digital media, in particular the capacity of accessing and hyperlinking information over computer networks and the Internet, also relies on electricity. With digital technology and the Internet, humans have come close to simulating the dynamics of thinking itself, in particular the capacity for recalling and cross-referencing data—hyperlinking—and through this process turning data into meaningful information. Given that digital media in its operation mimics thinking process of human mind or is an extension of our nervous system (brain included), as Marshall McLuhan would say, it is appropriate to take some liberty with language and use the term hyperlink in connection with human thought and its many other extensions, i.e., other media of communication.

With written communication, a non-digital form of hypertexting arose in handwritten manuscripts in terms of marginalia and illuminations, and later in printed documents including books, magazines, and newspapers. We explore these forms of hyperlinking as well as the digital variety that arose with digital computers, computer networks, and the Internet, through the introduction of which hyperlinking became standard mode of navigating the web, as well as hypermedia associated with nonlinear and interactive media such as the World Wide Web and standalone applications on mobiles for interactive books, videos, music, etc.

In Chapter 11, we focus on the spiral, cyclic, and hyperlinked forms in the arts including literature, poetry, painting, sculpture, architecture, music, and dance. Examples include Petrarchan sonnets, Dante's Divine Comedy, the literature of Proust and James Joyce's Finnegans wake, the art movement of Vorticism, the music of Bach and the fugues of other composers, cinema such as Sorel Etrog's film *Spiral* and Marcel Duchamp's *Anémic Cinéma*, the architectural form of the golden spiral and the spiral staircase, and the dance of Loïe Fuller and Martha Graham. We also consider the linking of information from one domain to another in the general biosphere with organisms other than humans to once again illustrate the universality of this topological structure as we did for the spiral structure.

The nonlinear linking and hyperlinking structures that we consider in this study are rhizomatic. The etymology of rhizome is from the Greek rhizome, which means "mass of roots." Hyperlinks on the web are like a mass of roots that branch out like in an iris flower or turmeric, which has a root system consisting of multiple nodes from which roots and shoots grow. Rhizomatic forms, or forms created by navigating through hyperlinks, have multiple entry and exit points, defy linearity, and are non-hierarchical.

The study of spiral, cyclic, and hyperlinked [rhizomatic] structures in the natural world perhaps represents a deviation from the guidelines of the series Mathematics in Mind to which this book belongs as the main focus of our study is the topology of the human mind, but we introduce these topics because they demonstrate the universality of spiral, cyclic, and hyperlinked structures. It also might help us understand why the language of mathematics is such a powerful and useful tool for describing the many levels of organization throughout our vast and varied universe.

The reader might wonder why we examine spiral structures in both the abiotic and biotic worlds but only touch upon hyperlinked [rhizomatic] structures in the biotic world and do not treat hyperlinking in the abiotic world. The reason is that we believe that information has no meaning as far as the behavior of abiotic matter is concerned as abiotic matter makes no choices as is the case with living organisms. Abiotic matter blindly follows the laws of nature or physics. Information as defined by Douglas MacKay (1969) is "a distinction that makes a difference" and redefined a short time later by Gregory Bates (1973) as follows "what we mean by information—the elementary unit of information—is a difference which makes a difference." Only a living organism can make a distinction or difference that makes a difference because they can make a choice as to how they behave and information is that which helps them make those choices. Abiotic matter cannot make any choices as any change in its state is governed by universal laws and hence it cannot be informed. Living organisms are informed by the information that they interpret and which they transform into a meaningful action or response.

We remind the reader that the origin of the term information comes from the notion of forming the mind. The English word information according to the Oxford English Dictionary (OED) first appears in the written record in 1386 by Chaucer: "Whanne Melibee hadde herd the grete skiles and resons of Dame Prudence, and hire wise informacions and techynges." The word is derived from Latin through French by combining the word inform meaning giving a form to the mind with the ending "ation" denoting a noun of action. The word makes its way into Middle English enforme, informe "give form or shape to," also "form the mind of, teach," from Old French enfourmer, from Latin informare "shape, fashion, describe." These earliest definitions refer to an item of training or molding of the mind. The next notion of information, namely the communication of knowledge appears shortly thereafter in 1450. "Lydg. & Burgh *Secrees* 1695 Ferthere to geve the Enformacioun, Of mustard whyte the seed is profitable."

This does not mean there is no information associated with abiotic matter. But that is not information that abiotic matter possesses but rather it is the information that we humans have about abiotic matter. After all, the natural sciences represent our information about and knowledge of abiotic matter. But the subjects of natural science, abiotic matter, themselves have no interaction with information and do not possess information.

Since only living organisms can be informed because they have choice as to how to realize their purpose or telos to propagate their organization (Kauffman et al. 2007), only they can transform stimuli or data into information and be influenced by information. When natural scientists talk about the exchange of information among different forms of abiotic matter they are speaking metaphorically. What about quantum information you might ask? The atomic states in which information is inscribed has no more information than the ink used to print this book. The atomic states and the ink used to print this book do not possess any information; they are not informed; they are just two different inanimate objects or media used to represent information. The same can be said for the spiral structures or forms and hyperlinks [rhizomes] that will be the subjects of our study. They do not possess information; they merely represent information. There was a folk belief in the Middle Ages that if one could wash the ink of a handwritten manuscript and drink it, one would possess the information of that manuscript. Qubits and ink do not possess information; they merely represent information.

Responding to the Topics and Questions Raised by the Series Editor

In developing our thoughts on a topology of mind, we will also specifically address many of the topics and questions raised by the Series Editor, Marcel Danesi in his call for proposals for this series. The topics and questions that we have selected to address in this study include the following that Professor Danesi suggested. We also have indicated how we plan to address each of these topics and questions:

- The historical context of any topic that involves how mathematical thinking emerged, focusing on archeological and philological evidence.
- Connection between math cognition and symbolism, annotation, and other semiotic processes.
- Interrelationships between mathematical discovery and cultural processes, including technological systems that guide the thrust of cognitive and social evolution.
- Is mathematics an innate faculty or is it forged in cultural-historical context?

We will make use of the archeological evidence of Denise Schmandt-Besserat (1978) who showed how mathematical notation for numbers first emerged in Mesopotamia when clay tokens used as receipts for tributes given to priests were embedded in clay tablets to reveal the first examples of mathematical notation and thinking. The first mathematicians were accountants. Schmandt-Besserat (1978) also showed that the emergence of mathematical notation occurred simultaneously with the emergence of writing and that symbols representing numbers and words appear together on the same tablets. This story will reveal the connection between math cognition and the semiotic processes of symbolism and annotation and at the same time shows the interrelationships between mathematical discovery and cultural processes, including technological systems that guide the thrust of cognitive and social evolution. The close connection of mathematics and writing has implication for mathematics education, especially for those who learn arithmetic first.

The story of the origin of mathematical notation and writing will also allow us to conclude that mathematics, other than simple enumeration, is forged in the cultural-historical context and is not an innate faculty, given that mathematics thrives in cultures that possess written language.

The next batch of questions we will address and how we will address them are the following:

- Other thematic areas that have implications for the study of math and mind, including ideas from disciplines such as philosophy, linguistics, and so on.
- Is mathematics a unique type of human conceptual system, sustained by specific and localized neural structures, or does it share neural systems with other faculties such as language and drawing?
- What structures, if any, do mathematics and language share?

These next three items will be addressed when we show how the emergence of spoken language led to the kind of conceptualization that makes mathematical thinking possible. Here, we will make use of Logan's (2007) study *The Extended Mind: The Emergence of Language, the Human Mind and Culture* where it is shown that the emergence of spoken language allowed humans to create the conceptualizations necessary for mathematical thinking beginning with simple enumeration. Before humans had language, the brain was basically a percept processor that did not have access to concepts. Our first words were also our first concepts. The word water was a concept that united all of our percepts with the water we drink, cook with, wash with, falls as rain, and which we find in rivers, lakes, and oceans. Without language we cannot conceive of quantities, the basic atoms of mathematics. With verbal language, the hominid brain bifurcated into the human mind fully capable of the conceptualization that made mathematical thinking possible. This idea is represented by the mathematical formula: mind = brain + language.

Spoken language gave rise to written language which allowed more complex forms of mathematical thinking beyond enumeration to develop. This establishes that mathematics shares neural faculties such as language and graphic representation as mathematical notation arose out of the construction of three-dimensional clay tokens followed by them being embedded in clay tablets as explained by Denise Schmandt-Besserat (1978). Each token represented a different agricultural commodity. The tokens for the large and small measures of grain came to represent the numbers 10 and 1, respectively. To distinguish the visual signs representing the numbers 10 and 1 from the signs representing the words for the large and small measures of grain, the signs representing words were not created by pushing the clay tokens into the wet clay, but rather the shape that the tokens representing agricultural commodities were drawn using a stylus work on the wet clay. The signs representing the numbers 10 and 1 continued to be created by pushing the tokens into the wet clay. As a result, the first notational system for representing words and numerals emerged at the same time some 5000 years ago with a separate notation for words created by sketching with a stylus and numerals created by imprinting the tokens into the wet clay.

How This Book Came About as the Result of Many Collaborations

This book emerged from three sources:

1. A collaboration dating back to 2011 on spirals and cyclic processes initiated by Izabella Pruska-Oldenhof's observation of the existence of a spiral structure in the thinking of Marshall McLuhan and a number of other thinkers, an insight that she shared with Bob Logan.
2. A collaboration that was initiated by the University of Toronto undergraduate student Emma Findlay-White, during an independent study supervised by Bob

Logan. This study revealed that hyperlinking is not just a recent phenomenon that emerged with the Internet and hypertext but that it represents a way in which the human mind engages in and organizes its thoughts and communications dating back to the oral composition of epic poetry such as Homer and to the organization of the first printed form of the Talmud by Joshua Solomon Soncino in 1483 more than 500 years before the appearance of the World Wide Web. We also developed the idea that the very origin of spoken language can be thought of as a result of the hyperlinking of concepts with percepts and the origin of writing and mathematical notation with the hyperlinking of visual signs with elements of spoken language. Emma's research during her independent study was a significant inspiration for parts of this book and her collaborations and contributions are hereby acknowledged.

3. Some earlier studies of Robert K. Logan published in the journal *Semiotic* and three of his books:

   - *Semiotica* 125-1/3 (Logan 1999) "The Social, Economic and Educational Impacts of Notational Systems"
   - *Semiotica* 155-1 (Logan and Schumann 2005) "The Symbolosphere, Conceptualization, Language and Neo-Duality" co-authored with John Schumann
   - *Semiotica* 160 (Logan 2006) "Neo-dualism and the Bifurcation of the Symbolosphere into the Mediasphere and the Human Mind"
   - *The Alphabet Effect*: Logan (2004a)
   - *The Sixth Language: Learning a Living in the Internet Age* (Logan 2004b)
   - *The Extended Mind: The Emergence of Language, the Human Mind and Culture* (Logan 2007)

The material in this book draws upon a number of other previous collaborations including a collaboration of Logan (2004a) with Marshall McLuhan in which a link was found between phonetic writing and the alphabet in particular with codified law, monotheism, abstract science, deductive logic, and hence axiomatic geometry.

A second collaboration between Logan and Denise Schmandt-Besserat, who discovered the link between the written word and mathematical notation through the interpretations of her archeological finds in the Near East. This collaboration led to a better understanding of mathematics learning and the evolution of language from speech to writing and mathematics, to science, to computing, and to the Internet as described by Logan (2004b) in the book *The Sixth Language: Learning a Living in the Internet Age*.

This study also draws upon the ideas developed in *The Extended Mind: The Emergence of Language, the Human Mind and Culture* (Logan 2007) which developed as a result of Logan's interaction with two communities, one with linguists and the other with emergent dynamicists or complexity theorists and therefore draws upon the ideas and suggestions of the following scholars Morten Christiansen (1994) and Terrence Deacon (1997, 2012) from linguistics and emergent dynamics; and Ilya Prigogine (1980) and Stuart Kaufmann (1995) from complexity theory and emergent dynamics.

This book, like all the other books that have ever been written, is not complete but at some point, one has to report one's findings. We invite you, the reader to fill in the gaps, the things that we missed, because we believe that mathematics and specifically spirals and hyperlinking are a fundamental part of human cognition. We invite you, the reader, to fill in the gaps and if you would be so kind to share them with us at the following e-mail addresses i2pruska@ryerson.ca and logan@physics.utoronto.ca.

Toronto, ON, Canada

Robert K. Logan
Izabella Pruska-Oldenhof

References

MacKay, Douglas. 1969. *Information, Mechanism and Meaning*. Cambridge MA: MIT Press.
Bateson, Gregory. 1973. *Steps to an Ecology of Mind*. St. Albans: Paladin Frogmore.
Kaufmann, Stuart. 1995. *At Home in the Universe*. Oxford UK: Oxford University Press.
Kauffman, Stuart, Robert K. Logan, Robert Este, Randy Goebel, David Hobill and Ilya Shmulevich. 2007. "Propagating organization: an enquiry." Biology and Philosophy 23: 27–45.
Schmandt-Besserat, Denise. 1978. "*The Earliest Precursor of Writing*". Scientific American 238.
Logan, Robert K. 2007. *The Extended Mind: The Emergence of Language, the Human Mind and Culture*. Toronto: University of Toronto Press.
Logan, Robert K. 1999. "The Social, Economic and Educational Impacts of Notational Systems." *Semiotica* 125-1/3: 15–20.
Logan, Robert K. and John H. Schumann. 2005. The symbolosphere, conceptualization, language and neo-dualism. *Semiotica* 155 (1), pp. 201–14.
Logan, Robert K. 2006. "Neo-dualism and the Bifurcation of the Symbolosphere into the Mediasphere and the Human Mind." *Semiotica* 160; 229–42.
Logan, Robert K. 2004a. *The Alphabet Effect: A Media Ecology Understanding of the Making of Western Civilization*. Cresskill NJ: Hampton Press (1st edition 1986. New York: Wm. Morrow).
Logan, Robert K. 2004b. *The Sixth Language: Learning a Living in the Internet Age*. Caldwell NJ: Blackburn Press (1st edition 2000. Toronto: Stoddart Publishing; Mandarin Edition ISBN 986-7964-05-5).
Christiansen, Morten. 1994. Infinite languages finite minds: Connectionism, learning and linguistic structure. Unpublished doctoral dissertation, Centre for Cognitive Studies, University of Edinburgh UK.
Deacon, Terrence. 1997. *The Symbolic Species: The Co-Evolution of the Brain and Language*. New York: W. W. Norton & Co.
Deacon, Terrence. 2012. *Incomplete Nature: How Mind Emerged from Matter*. New York: W. W. Norton and Company.
Prigogine, Ilya. 1980. *From Being to Becoming*. New York: W. H. Freeman and Company.

Acknowledgments

We want to thank Professor Marcel Danesi for the invitation to publish our results in this Springer series Mathematics in Mind.

Contents

Prologue to Part I Mathematics in Mind

We will develop the hypothesis in Part I—Mathematics in Mind that the human mind is intrinsically verbal and mathematical and that language and mathematical thinking co-emerged at the dawn of the emergence of Homo sapiens. Language and mathematical thinking are the difference that separates human beings from the rest of living beings including our most recent ancestors who also belong to the genus Homo ranging from Homo habilis and Homo erectus to Homo Neaderthalensis and Homo Denisova (the jury is out as to whether denisovans are a sub-species of sapiens or a separate species).

In Chap. 1 we scope out the nature of mathematics in the human mind, which we show involves much more than just the formal study of mathematics in schools and the academic discipline of professional mathematicians. All human thought is at a certain level mathematical. We describe the different ways that mathematics enters human thought and language.

In Chap. 2 we will develop the hypothesis as an abduction that (1) verbal language, which permitted conceptualization, and (2) a primitive form of set theory or the ability to classify percepts or experiences, co-emerged creating the human mind and the fully human species Homo sapiens. In other we will argue that the ability to speak and mathematics in the mind in the form of set theory originated simultaneously. We also argue that language is not just the medium to communicate thought but that it is the medium in which thought emerges. The association of a word as a concept with a set of percepts related to that word/concept has a mathematical structure to it. Finally, we describe the emergence of grammar.

In Chap. 3 we study the impact of language both oral and written on the formulation of thought. We then examine the impact of the emergence of writing and mathematical notation on mathematical thinking. We will show that writing to express human thoughts and mathematical notation to represent quantities arose simultaneously. We trace the historic origin of the concept of zero and the following mathematical concepts that followed in its wake including the place number system, negative numbers, infinity, and algebra. In particular we examine the effect of phonetic writing systems including the alphabet and their effect on and/or relation to codified law, monotheism, deductive logic and abstract science. We describe the evolution of language from speech into writing, mathematics, science, computing and the Internet.

In Chap. 4 we study the impact of alphabetic writing on the ancients Greeks on their development of deductive logic and its application to geometry and abstract science. We describe the difference between deductive logic and abstract science and demonstrate that one cannot prove the truth of a proposition using the methods of science.

In Chap. 5 we review the evolutionary chain of the six languages of speech, writing, mathematics, science, computing and the Internet. We then show that the information overload created by science led to the emergence of computing and that the information overload created by computing led to the emergence of the Internet and the Web. We then describe the semantical and syntactical aspects of the fifth and sixth languages of computing and the Internet/Web. We conclude this chapter with a discussion of computer-based AI and show how great is the difference between human intelligence and artificial intelligence debunking the notion of the technological Singularity.

Chapter 1
The Origin of Mathematics in the Mind

Introduction

Before starting our study of the topology of the thought and communications that the human mind creates as part of the **Mathematics in Mind** series we asked ourselves the question: what is the origin of mathematics in mind? To answer this question, we must first ask, what is the origin of the mind and what is the origin of mathematics. We will start first with the question of what is mathematics in this chapter and then turn to the question of the origin of the mind in the next chapter. We will discover that one of the aspects of mathematics, namely set theory, played a significant role in the creation of verbal language and, subsequently, the human mind and propose that set theory and verbal language co-emerged. The human mind as we will show in the next chapter is a product of the brain and verbal language or put another way the verbal language extended the brain into the mind. But for now, let us explore what is the nature of mathematics.

What Is Mathematics?

Mathematics is many different things depending on who is defining it because it has so many different facets. Mathematics is concerned with quantity, number, enumeration, magnitude, order, arrangement, structure, form, shape, space, change, calculation, and measurement. There is in fact no simple definition of mathematics so rather than trying to define mathematics in a simple sentence we will merely describe the many forms that mathematics takes. The term mathematics etymologically speaking is derived from the ancient Greek term μάθημα (máthēma), which translates into English as knowledge, learning or study and hence describes a discipline. But long before the ancient Greeks began the formal study of mathematics, the Babylonian and the Egyptians also studied math as a discipline as evidenced by

R. K. Logan, I. Pruska-Oldenhof, *A Topology of Mind*, Mathematics in Mind,
https://doi.org/10.1007/978-3-030-96436-8_1

their respective texts Plimpton 322 (Freiberg 1981) and the Rhind Mathematical Papyrus (Neugebauer 1969) both circa 1900 BC. The appearance of mathematics, not as a discipline, but as a way of thinking or as a cognitive structure can be traced back to the very beginning of human existence and the simultaneous emergence of verbal language and the human mind as we will describe when we discuss the idea of the extended mind. Mathematical thinking is a universal characteristic of all human cultures as is evidenced by the fact that all languages have words for numbers and other ways of designating amounts and orderings. Mathematical thinking seems to have been a characteristic of humankind throughout our history. One piece of evidence for this is the archeological find of the 22,000-year-old Ishago Bone on which a numerical tally was etched (De Heinzelin de Braucourt 1962).

The level of mathematical activity varies from culture to culture. Hunting and gathering societies, for example, did not have much need for mathematical thinking and as a result the range of numerals in these societies is very simple consisting of the numbers one, two, and many (Boyer 1991, 3). Even though this is the case an experiment conducted by Brian Butterworth of University College London with Australian indigenous children revealed that they are able to distinguish numbers greater than two even though their number systems only designate 1, 2 and many. As a result of his study he remarked:

> Recently, an extreme form of linguistic determinism has been revived which claims that counting words are needed for children to develop concepts of numbers above three. That is, to possess the concept of 'five' you need a word for five. Evidence from children in numerate societies, but also from Amazonian adults whose language does not contain counting words, has been used to support this claim. However, our study of aboriginal children suggests that we have an innate system for recognizing and representing numerosities—the number of objects in a set—and that the lack of a number vocabulary should not prevent us from doing numerical tasks that do not require number words (https://www.ucl.ac.uk/news/news-articles/news-releases-archive/aboriginal).

This result should come as no surprise given that there are many animals capable of numeracy including lions, monkeys, certain birds, and chimpanzees. Cultures that required keeping track of quantities developed more sophisticated number systems. It seems as a result of Brian Butterworth's finding that numeracy is a built-in feature of humans.

In terms of cultures that have limited number words we found the following example from the Gumulgal people of Australia ingenious. They developed a number system base two with just two number words used repetitively. The word urapon represents 1 and the word ukasar represents 2, but they are able to represent numbers greater than two as follows: ukasar-urapon = 3; ukasar-ukasar = 4; ukasar-ukasar-urapon = 5 and ukasar-ukasar-ukasar = 6 and so on and so forth. Odd numbers end in urapon and even numbers in ukasar (https://numberwarrior.wordpress.com/2010/07/30/is-one-two-many-a-myth, accessed February 22, 2017).

Number words are only one indicator of mathematical thinking. Mathematics is concerned with the comparisons of quantity, amounts and number of items, on the one hand, and order or structure, on the other hand. Number words, whether cardinal or ordinal are a precise way of making these comparisons, but there are many other words that can also be used to make such comparisons and are evidence of mathematics in mind. Here is a list of such words in English that have their counterpart in many other languages.

Table of Words Related to Quantitative Analyses or Comparisons

Many, much, none, amount;
Some, more, most, almost;
Less, lesser, least;
Little, littler, littlest,
Tiny, tinier, tiniest;
Small, smaller, smallest;
Big, bigger, biggest;
Large, larger, largest;
Great, greater, greatest;
Far, farther, farthest, further, away;
Close, closer, closest, near;
Beside, between, behind, in front of, after, under, over, above;
Before (position and time);
After (position and time);
Long, longer, longest (position and time);
Short, shorter, shortest (short can refer to height but also not having enough);
Heavy, heavier, heaviest (weight but also an emotion);
Light, lighter, lightest (weight, tone and emotion);
Equal, unequal, approximately;
Extreme, slight, intense;
Now, later, latest, past, present, future, before (time), after (time);
Soon, sooner, soonest;
Dawn, sunrise, morning, noon, afternoon, evening, sunset, night, midnight;
Start, begin, during, end, finish;
When, where, how many, how much, how often;
Seldom, often, never;
Open, closed;

We have also compiled a separate list of words that describe mathematical operations that are also used outside the formal study of the discipline of mathematics that have meaning in addition to their formal mathematical definition. These etymologies reveal that these mathematical operations have their roots in the cognitive structures of the mind and that formal mathematics has its roots in the linguistic structures of the mind that emerged with verbal language, a topic we will return to in the next chapter.

Table of Mathematical Terms and Their Etymology That Also Have Significance Outside of Mathematical Studies

Add; from ad + dare ["to give, put"] (www.dictionary.com/browse/**adding**); to join or unite (something to something else), from Latin addere "**add** to, join, attach, place upon" (www.dictionary.com/browse/**adds**).

Take away or subtract; 1530s, "withdraw, withhold, take away, deduct," a back-formation from **subtraction** (q.v.), or else from Latin subtractus, past participle of subtrahere "take away, draw off." Related: **Subtracted**; **subtracting**. Mathematical calculation sense is from 1550s (www.etymonline.com/index.php?term=**subtract**).

Multiply; From multus ["much, many"] + plicō ["fold, double up"] (math.stackexchange.com/questions/1150438/the-word-times-for-**multiplication**); Online **Etymology** Dictionary. mid-12c., multeplier, "to cause to become many," from Old French multiplier, mouteplier (12c.) "increase, get bigger; flourish; breed; extend, enrich," from Latin multiplicare "to increase," from multiplex (genitive multiplicis) "having many folds, many times as great in number" (www.etymonline.com/index.php?term=**multiply**).

Divide; Online **Etymology** Dictionary. early 14c., from Latin dividere "to force apart, cleave, distribute," from dis- "apart" (see dis-) + -videre "to separate," from PIE root *weidh- "to separate" (www.etymonline.com/index.php?term=**divide**).

Equal; late 14c., "identical in amount, extent, or portion;" early 15c., "even or smooth of surface," from Latin aequalis "uniform, identical, **equal**," from aequus "level, even, flat; as tall as, on a level with; friendly, kind, just, fair, equitable, impartial; proportionate; calm, tranquil," which is of unknown origin (www.etymonline.com/index.php?term=**equal**).

These etymologies reveal that these mathematical operations have their roots in the cognitive structures of the mind and that formal mathematics has its roots in the linguistic structures of the mind that emerged with verbal language, a topic we will return to in the next chapter.

Although mathematics is a discipline that is studied for its own sake mathematical thinking is an integral part of human cognitive structures and many forms of human activity. Enumeration or counting and the subsequent practice of arithmetic including adding, subtracting, multiplying and division are an essential part of commerce. Another form of mathematical thinking entails the measurement of amounts by volume or by weight; distances between points and land areas, all of which are essential for the organization of agriculture and the commerce associated with it, which resulted in standard measures of volumes, weights and distances and in the case of land areas to geometry.

Organization and Classification as a Form of Mathematical Thinking

Mathematics is primarily about quantitative analysis but it is also about qualitative analysis and organization as well. The organizing of objects and/or concepts in groups, sets, classes, classifications, categories, taxonomies, or schemas is an intrinsically mathematical operation. In a certain sense qualitative analysis comes before quantitative analysis because in order to enumerate or count things one has to first determine if the objects being counted or enumerated belong to the same class or group. Before one can say whether there are five dogs one must first determine that the objects being counted are in fact dogs. So, classification must precede enumeration. We therefore conclude that the ability to group objects in categories or classes is basically a mathematical skill. We will make use of this notion when we come to understand how the emergence of verbal language by the first Homo sapiens is connected to the mathematics in mind of connecting a group or set of percepts or experiences to a single concept represented by a word. Here we are using the terms group and set not as they are defined in group theory and set theory respectively but rather as a collection of percepts that are similar. We are suggesting that a primitive form of set theory and verbal language co-emerged.

Mathematics: A Practical Tool and a Theoretical Discipline

The etymology of the word mathematics from the Greek word for study or knowledge reveals that mathematics is considered as both a discipline of study for its own sake for scholars as well as a subject to be learned in school for practical reasons as mathematics is an integral part of many of the practical aspects of human life including the day to day organization of daily life as well as the following human activities that are unique to our species, namely, agriculture, trade, commerce, accounting, statistics, probability, insurance, business, management, finance, construction, engineering, science, medicine, computing, government, politics, and music. This list is by no means complete but it does indicate the diversity of activities in which mathematics plays a role.

The Different Divisions and Sub-disciplines of Mathematics

The field of mathematics includes many different divisions and subdivisions, which we will enumerate in roughly the order in which they emerged historically. The most basic form of mathematics is simply counting or enumeration. Next comes the simple arithmetic of adding, subtracting, multiplying and dividing whole numbers, fractions and decimals (actually the use of decimals are only 500 years old whereas fractions date back almost 4000 years to Egypt and Babylon). There then follows geometry in both two and three dimensions followed closely by syllogistic logic. The next development is that of the concept of zero, negative numbers and algebra developed in India approximately 2000 years ago and transmitted to Europe in the fourteenth century by Arab mathematicians; followed by the notion of square roots, quadratic equations and raising numbers to exponential values. The Renaissance also saw establishment of double entry bookkeeping. Next came analytic geometry which gave rise to calculus in the seventeenth century as well as the introduction of logarithms and probability theory. The eighteenth century ushered in the use of imaginary numbers, graph theory, topology, combinatorics, statistics and number theory. The nineteenth century saw the development of functions of complex variables, non-Euclidean geometry, vector spaces, noncommutative algebra, Boolean algebra, group theory, set theory, symmetry studies, and mathematical logic. The twentieth century saw the development of differential geometry, game theory, category theory, ergodic theory, knot theory, catastrophe theory, fractals, emergent dynamics, complexity theory, Lie theory, Turing machines, information theory, and the incompleteness theorem.

Chapter 2
Mathematics, the Human Mind, Verbal Language: Mathematics in Mind

Introduction

In Chap. 1 we scoped out the domain of mathematics. In this chapter, we explore the complex relationship between mathematics, spoken language, written language, mathematical notation and mathematics as a language. We will also explore what is meant by the phrase 'mathematics in mind', which is the theme of the series to which this book belongs. To this end we will develop our notion of what is meant by mind and show that it is more than just the human brain but that it is intimately connected to language. We will argue that not only is mathematics a language but that one aspect of mathematical thinking contributed to the emergence of verbal language and that in turn led to the bifurcation of the brain into the mind capable of conceptualization and the development of mathematics as we know it today. We will propose that the origin of verbal language, the origin of the mind and the origin of mathematic thinking all happened at approximately the same time and that these three elements are basically hyperlinked. That is our bold hypothesis that we will develop in this chapter. In addition to this we will also examine the role that written language and hence mathematical notation played in the development of mathematical thinking and practice.

The Origin of Mind and the Role of Mathematical Thinking

In order to discuss mathematics in mind we first need to understand what we mean by mind, which is not the same as the human brain and, as we will see, is intimately connected to the origin of verbal language which in turn is connected to mathematical thinking in terms of classification. In order to develop this hypothesis, we will need to make use of abductive reasoning also known as an abduction, which is a fancy way of saying a just-so story. Just to remind the reader an abduction is a

R. K. Logan, I. Pruska-Oldenhof, *A Topology of Mind*, Mathematics in Mind,
https://doi.org/10.1007/978-3-030-96436-8_2

proposed explanation of a phenomena formulated as a hypothesis that potentially explains that phenomena and is not necessarily true but is worthy of consideration and further study. As described in the Wikipedia article on abduction (https://en.wikipedia.org/wiki/Abductive_reasoning, accessed on January 21, 2017),

> Abductive reasoning is a form of logical inference which goes from an observation to a theory which accounts for the observation, ideally seeking to find the simplest and most likely explanation. In abductive reasoning, unlike in deductive reasoning, the premises do not guarantee the conclusion. One can understand abductive reasoning as "inference to the best explanation."

The other forms of reasoning are deductive and inductive. Deductive reasoning starts with a set of axioms that are so obviously true that no one could possibly dispute them. From these axioms one proceeds to draw conclusions from these self-evident axioms. There are no such self-evident truths about the nature of mind, language or mathematics as there are many conflicting views on the nature of these phenomena. Inductive reasoning is out of the question as there are not many examples as there is only one species of living organisms, humans, that has a mind that engages in verbal language and mathematics notwithstanding the fact that animals communicate with each other and that there are some animals who have a sense of numbers including lions, some birds and some apes.

So, we are stuck with abductive reasoning to address the question of the origin of mathematics in mind and the origin of verbal language and the origin of the mind itself. As an abduction starts with an observation, our abductive process starts with the observation that human beings are the only animals that communicate with verbal language and we are also the only animal capable of abstract thinking, planning and the ability to deal with situations that are not exclusively in the here and now. In other words, we can conceive of a past, present and future. Birds build nests, beavers build dams and squirrels bury nuts but these activities are instinctive and not the result of abstract thinking about the future. Non-human animals solve problems in the here and now but only humans can think about and communicate about events in the past and events yet to come.

Before embarking on this study, we want the reader to understand that we will make a distinction between the brain and the mind. All vertebrates have a brain but humans are the only species that have a mind that evolved from their brain that has the added features of verbal language, mathematical thinking, the ability to plan and the ability to conceive of things that are not immediately available to them in the here and now. We also posit that thought, language and communication are interconnected or hyperlinked. Many linguists regard language as the medium by which we communicate our thoughts. We will develop the hypothesis that language is the medium in which we formulate our conceptual thinking. We see thinking as silent language and that language also has the additional feature of facilitating communication. Thinking, communicating and language form an emergent supervenient system.

Language Facilitates Thinking and Is the Medium in Which Thought Is Conceived

Language in not just our medium of communication but it is also the medium in which we formulate our thoughts. Some will claim that they have thoughts that are non-verbal as they are able to conjure up images and feel pleasures that are non-verbal. What people call non-verbal thinking is basically their perception of sensations and sensations are not something that are verbal, they are merely percepts. The core of the idea that we are developing is that percepts and concepts are very different. Percepts are non-verbal sensations that are very hard to put into words because they originate in the sense organs of our body whereas concepts are the products of our mind and the thoughts that our mind produces. Thoughts are conceived in terms of words and are therefore easy to communicate. Thought and verbal language are intimately connected. We see thinking as a form of silent language. We are not alone in this.

Darwin's (1871) expression of the co-evolution of language and the intellectual power of humans can be found in Chap. 21, p. 92 of The Descent of Man: "A complex train of thought can no more be carried on without the aid of words, whether spoken or silent, than a long calculation without the use of figures or algebra."

Deacon (1997, 22) also sees cognition or symbolic thought as one of the main functions of speech,

> Language is not merely a mode of communication, it is also the outward expression of an unusual mode of thought—symbolic representation.... Symbolic thought does not come innately built in, but develops by internalizing the symbolic process that underlies language.

Language is not a passive container whose only function is to transmit and communicate our ideas and sentiments from one person to another. Language is the medium for the actual conceptualization of thought. Language is a "living vortice of power" (McLuhan 1972a, v) that shapes and transforms our thinking. Language is both a system of communication and an informatic tool as McLuhan (1964, 5) observed:

> All media are active metaphors in their power to translate experience into new forms. The spoken word was the first technology by which man was able to let go of his environment in order to grasp it in a new way. Words are a kind of information retrieval that can range over the total environment and experience at high speed. Words are complex systems of metaphors and symbols that translate experience into our uttered or outered expression. They are a technology of explicitness. By means of translation of immediate sense experience into vocal symbols the entire world can be evoked and retrieved at any instant.

Language not only allows humans to communicate abstract concepts to each other but it is also put to use for the internal dialogue of conceptual thinking. The two activities of communication and thinking emerged simultaneously. The creation of verbal language represents a bifurcation from concrete perceptual mental activity to the creation of abstract concepts in thought and communication. Verbal language was a classic example of punctuated evolution in which the non-verbal, mimetic, percept-based preverbal forms of language of pre-human hominids evolved into verbal concept-based speech and the inner voice which we call conscious thinking. Merlin Donald (1991) suggested that mimetic communication consisting of hand signals, facial gestures, body language and pre-verbal prosody or tone such as grunts was the cognitive laboratory in which verbal language emerged.

Children ages 3–7 resort to egocentric speech when-ever they are confronted with a puzzling situation that they need to think through. Vygotsky (1962) suggested that that egocentric speech is basically the child thinking out loud, and that it naturally evolves into inner speech once the child realizes that the vocalization is not necessary for the main function of this form of speech, namely, problem solving or thinking. "We came to the conclusion that inner speech develops through a slow accumulation of functional and structural changes, that it branches off from the child's external speech simultaneously with the differentiation of the social and the egocentric functions of speech, and finally that the speech structures mastered by the child become the basic structures of his thinking (ibid., p. 51)."

Vygotsky's model is based on the notion that speech or language has two components: communications and informatics. The language of children younger than age 3 is social speech and is purely about communication and social interaction. With egocentric and inner speech, language becomes a tool to assist the child to think and process information. The only difference between egocentric and inner speech is that the former, which appears first for processing information, is vocalized, and the latter is not. But both serve the same function. Egocentric speech disappears and vocalized speech is used exclusively for communication purposes once inner speech emerges and is used solely for the purposes of thought. To conclude we suggest that rather than regarding speech as vocalized thought one may just as well regard thought as silent speech.

Language Extends the Brain into a Mind

Our story begins with the study, *The Extended Mind: The Emergence of Language, the Human Mind and Culture* (Logan 2007) which posits the simultaneous origin of verbal language and the emergence of the mind as a bifurcation of the brain as a percept processor into a mind capable of conceptualization so that the mind = brain + language. It was while considering the origin of mathematical thinking that we realized that the Extended Mind story also has a connection with the emergence of mathematics in mind in the form of classification or grouping like things into sets or groups and giving a name to that set or group. It is the same skill that eventually gave rise to set theory and group theory and as we will argue preceded the skill of counting or enumeration but actually laid the foundation for it.

We believe that the human mind emerged from the hominid brain at the same time verbal spoken language emerged. We have also argued that our first words were concepts that linked to and represented the percepts associated with those words. We therefore suggest that the operation of the mind entails making links (or hyperlinks if you will) and associations between words as concepts and the percepts of the world that are sensed or perceived.

Words represent concepts and concepts are represented by words. It is our belief that they emerged simultaneously and that words provided a medium by which concepts could be represented, manipulated, spoken about and thought about. This differs dramatically from the position of traditional linguists who believe that words emerged for the purpose of the communication of concepts that already existed before language existed. In other words, we believe that conceptual thought was not possible without language. We share the view of certain linguists that claim that words and concepts are connected.

Pinker (2003) suggested that "a word is an arbitrary sign; a connection between a signal and a concept." Where we differ is on the question of which came first the word or the concept. For Pinker first comes the concept and then the word whereas we believe that they co-emerged. We believe that the word gave rise to the concept and the concept was represented by the word. The word is more than a symbol or a sign that represents a thing or a concept. To my way of thinking the word is the concept and the concept has a handle, which is the word. The origin of language and words is tied to the origin of concepts so to understand why language arose we need to understand why we needed concepts.

Because the concept has a handle in the form of a word its use links the many percepts the speaker has associated with the concept. The word therefore facilitates the speaker's memory of those percepts and the ability to use those past experiences to plan new ways to incorporate those past experiences into actions that enhance the speaker's survival. Words help organize the user's past experiences to make a better future.

The use of a word like water representing the concept of water triggers instantaneously all of the mind's direct experiences and perceptions of water such as the water we drink, the water we cook with, the water we wash with, the water that falls as rain or melts from snow and the water that is found in rivers, ponds, lakes, and oceans. The word water also brings to mind all the instances where the word "water" was used in any discourses in which that mind participated either as a speaker or a listener/reader. The word "water" acting as a concept and an attractor not only brings to mind all "water" transactions but it also provides a name or a handle for the concept of water, which makes it easier to access memories of water and share them with others or make plans about the use of water. Words representing concepts speed up reaction time and, hence, confer a selection advantage for their users. And at the same time those languages and those words within a language, which most easily capture memories enjoys a selection advantage over alternative languages and words respectively.

The Origin of Verbal Language: The Bifurcation from Percepts to Concepts

We have suggested that spoken language arose simultaneously with the transition from percept-based thinking to concept-based thinking with spoken words representing a concept linking the set of percepts associated with the concept represented by the word. One might ask the question what motivated this transition. We will suggest in this chapter that spoken language was the mechanism to deal with the information overload that arose with humans being able to systematically control fire as described by Logan (2007, 53–54). The motivation for this hypothesis came from a previous study, The Sixth Language (Logan 2004b) that described the evolution of spoken language into the additional five languages of writing, mathematics, science, computing and the Internet that arose in the order in which they are listed with the exception that writing and mathematical notation arose at the same time. In Chap. 3 we will show that each new language arose from the previous language or languages to deal with the information overload that those language(s) created and could not deal with. Logan (2007) in The Extended Mind proposed that spoken language evolved or emerged from mimetic communication due to the information overload that ensued when humans were able to control fire and changed their social structure as a result. Instead of living in nuclear family groups early humans started to live in extended family clans in order to maintain the fire or the hearth around which their new life was centered. Logan (2007) suggested, or perhaps we should say hypothesized or abducted, that verbal language arose to deal with the information overload of coordinating the activities of the many people living around the campfire.

We start with the notion that pre-human hominid communication and hence thought processes were percept-based. Merlin Donald (1991, 226) in his book The Origin of the Modern Mind and elsewhere argues that "*The principle of similarity that links mimetic actions and their referents is perceptual, and the basic communicative device is best described as implementable action metaphor* (Donald 1998, 61)."

Our hominid ancestors emerged in the savannas of Africa, where they acquired the new skills of tool making, the control of fire, group foraging, and coordinated hunting. Fire provided warmth, increased their food supply and kept predators at bay. Controlling fire resulted in a change of social structures from living in nuclear family groupings to living in clans of related nuclear families so that the life-giving fire could be maintained. This more complex form of social organization increased the complexity of hominid life and resulted in a form of information overload. Percept-based communication and the resulting percept-based thought alone could not provide a sufficient amount of abstraction to deal with the increased complexity of hominid existence. We believe that with the information overload and chaos that ensued with the new social living arrangement, a new abstract level of order emerged in the form of verbal language and conceptual thinking which possessed the requisite variety to deal with the new challenges of group living and the need for social coordination.

According to Ross Ashby's Law of Requisite Variety, "*a model system or controller can only model or control something to the extent that it has sufficient internal variety to represent it* (Heylighen and Joslyn 2001)." When the complexity of hominid life became so great that percept-based communication and thinking could no longer provide the requisite variety to model or regulate the challenges of day to day life a new level of order based on concepts and verbal language emerged. Percepts arise from our impressions of the external world which our senses apprehend. Whereas concepts are abstract ideas that result from the generalization of particular examples. Concepts unlike percepts allow one to deal with things that are remote in both the space and the time dimension. Our first words were concepts that allowed us to represent things that are remote is both space and time.

Percepts are specialized, concrete and tied to a single concrete event but concepts are abstract and generative. Concepts increase the variety with which the brain, now a mind, can model the external world. They can be applied to a variety of different situations or events. They can be combined with other concepts and percepts to increase the complexity of what can be modeled, which is not possible with percepts alone. That is why only human are able to use their symbols generatively to create new ideas and to plan future activities. Some enculturated animals are able to comprehend symbols but they cannot use these symbols generatively to express new ideas. They use these sign indexically in the Peircean sense rather than symbolically.

The Mechanism for the Transition from Percept-Base Thinking to Concept-Based Thinking

What, we may ask, was the mechanism that allowed the transition from percept-based thinking to concept-based thinking to take place? Having assumed that language is both a form of communication and an information processing system we believe that the emergence of speech represented the actual transition from percept-based thought to concept-based thought. The spoken word is the actual medium or mechanism by which concepts are represented or expressed. The relationship of spoken language and conceptual thought is not a simple linear causal one. Concepts did not give rise to language and language did not give rise to concepts, rather conceptualization and the spoken word emerged at exactly the same point in time creating the conditions for their mutual emergence. Language and conceptual thought catalyzed each other's development. They are autocatalytic and are the linked parts of a dynamic cognitive system, namely, the human mind.

Kauffman used the notion to of autocatalysis to explain the emergence of life: "*A living organism is a system of chemicals that has the capacity to catalyze its own reproduction* (Kaufman 1995, 49)." An autocatalytic set of chemicals is a set of organic molecules for which the production of each member of the set is catalyzed by some other member of the set so that the system possesses a "*self-maintaining and self-reproducing metabolism.*" The system, i.e. a living organism, must operate in the presence of a source of energy and the basic atoms needed to build the organic compounds it needs to carry out its metabolism. A key idea in Kauffman's approach is that the members of the autocatalytic set self-organize and, hence, bootstrap themselves into existence as a set with properties that none of the components that make up the set possess.

An autocatalytic process that catalyzes itself through positive feedback loops so that once it starts, even as a fluctuation, it will accelerate and build so that a new phenomenon emerges. The emergence of language and conceptual thought is an example of an autocatalytic process. A set of words working together create a structure of meaning and thought. Each word in an utterance shades the meaning of the next words and the thought they express. Words and thoughts are both catalysts and products of each other. Language and conceptual thought are emergent phenomena. They bootstrap themselves into existence.

Percepts, Concepts, Their Relationship and Mathematics in Mind

The use of a particular word transforms the brain from one state to another and substitutes a set of percepts with a concept representing those percepts. The next step is that brain is able to find connections with the different words that it creates and as a result language emerges with relationships between the various words that have been created from their respective sets of percepts. Those relationships among the words become the syntax or grammar of the language. Those relationships become a pattern that can be applied to new words as the vocabulary or semantics of the language increases. Each individual word is a strange attractor for all the percepts associated with the concept represented by that word. The attractor is a strange attractor because the meaning of a word never exactly repeats itself. The trajectories of a strange attractor never meet even though they come infinitesimally close to each other. It is the same with a word. The meaning of a word fluctuates about the strange attractor but it is never exactly the same because the context in which the word is being used is always different. The context includes the other words in the utterance, who made the utterance, the social context in which the utterance was made, and the medium in which the utterance was made.

Each word in a language packs a great deal of experience into a single utterance or sign. Millions of percepts of a linguistic community are boiled down by the language to a single word acting as a concept and a strange attractor for all those percepts. This is where the notion of mathematics in mind comes into play in the form of the ability to group a set of percepts and represent them as a word acting as a concept. We believe that this is the same skill that mathematicians used to develop set theory in the nineteenth century beginning with Georg Cantor (https://plato.stanford.edu/entries/set-theory/, accessed April 30, 2021).

Concepts are 'artificial or virtual percepts'—instead of bringing the mountain or the percept of the mountain directly to the mind the word mountain brings the mind to the mountain through the concept of the mountain. The association of a word acting as a concept with a set of percepts has a mathematical structure to it. In mathematics a function is a mapping between two sets of values. We think of language as a mapping from a set of percepts to the single value of the concept associated with those percepts as we illustrated above with the word/concept of water (Prof. Charles Hamaker of St. Mary's College in California, private communication).

We have suggested that the brain before verbal language was merely a percept processor and that afterwards it was able to conceptualize, i.e. operate with concepts. Each concept linked all the percepts associated with that concept. We conclude that the human mind naturally makes associations, creates categories or sets and hence has the natural mathematical structure of set theory. We further suggest that verbal language emerges as a primitive form of set theory in that a set of percepts that are associated with each other or are similar are linked together with a word acting as a concept that unites all the members of that set. In a certain sense the primitive form of set theory we just described seems to be a pre-condition for the emergence of verbal language. It is not possible to determine the causal linkage between the primitive form of set theory and verbal language. We posit that set theory, vocalization and verbal language co-emerged as part of a complex general system where the bottom-up, top-down and lateral causal interactions cannot be separated. The best we can say is that vocalization, set theory and verbal language formed an autocatalytic set.

The Emergence of Grammar or Syntax

So far, we have provided a model or an abduction for the emergence of words or semantics. But there is another dimension to language that is equally important, namely syntax or grammar. Grammar is a pattern in which words as concepts are combined to create meanings through sentences or multi-word utterances. One imagines that the first utterances were a single word that embraced a single concept. But one can further imagine that as the number of words increased the first speakers found they could expand what they could say and think by combining words. As certain patterns of the words developed in vocal language grammar or syntax emerged. At first, the pattern was no more complicated than subject and verb, but over time a generative grammar emerged. I concede that this is a very simplistic model for how generative grammar came about but is Chomsky's model that generative grammar emerged basically as the result of a genetic mutation any more sophisticated.

The Emergence of Enumeration

I contend that the emergence of set theory that co-emerged with verbal language preceded enumeration. Enumeration can involve concrete numbers or abstract numbers. Concrete numbers are tied to the objects being enumerated and were likely the first form of enumeration appearing before abstract numbers. Examples of concrete numbers include a pair of shoes, a yoke of oxen or a brace of partridges. Concrete numbers have meaning only as units of the commodity they are designating and enumerating. An abstract number like ‘two’ can be applied to any set of two objects. We contend that concrete numbers preceded abstract numbers like the concept of one, two, three or any member of the set of numerals.

Concrete numbers, such as “a brace of partridges” or “a yoke of oxen” cannot be used to designate “two” as an abstract number and then be used to enumerate other objects. A brace of sandals is meaningless; instead one must refer to them as a “pair of sandals,” that is, as a concrete number or as “two sandals” where “two” operates as an abstract number. We would surmise or speculate or abduct that at some point in the evolution of language one particular concrete number came to represent an abstract number. We can only guess as to how this happened but certainly it is the case that numbers in the form of numerals like one, two and three are basically concepts represented by words. The number two is associated with the perception of two people or two deer or two eggs or, even more relevantly, two fingers. It is no accident that the universal number system of most cultures is 10, the number of our fingers or in some case 20 where counting includes both the fingers the toes. The term digit has two meanings, one meaning is a finger and the other meaning is a numeral. This is not an accident.

Just as we suggested that the word water was used to represent the many percepts associated with water the words for numbers could have arisen in a similar manner. When confronted with two objects that were similar like two persons or two eggs the idea of two things that were alike could be represent by the numeral two and in a similar way for three of a kind, four of a kind, etc. etc. (This idea or abduction was developed with the help of Dr. Juan Mansilla during a Skype conversation on November 17, 2017).

The notion that set theory played a role in the emergence of words and language might also explain the minds ability to create metaphors. Consider a set of percepts that gives rise to a word. Then consider a second set of percepts that shares some properties with the first set and gives rise to a second word that is different than the first word. In this instant, a metaphor can be created for the second word using the first word because of the shared features of the two set of percepts. We therefore abduce that the ability to create sets allows the mind to create metaphors. The word metaphor derives from the Greek words ‘meta’ meaning across and ‘phorein’ to carry. The word metaphor literally means to carry across where meaning from one situation is carried across to another situation. The word metaphor is itself a metaphor of ‘carrying across’.

The use of words transforms the brain by replacing a set of percepts with a concept and as a result the mind = brain + language emerges. In other words, with language the brain bifurcates into the mind capable of conceptualization and abstract thinking such as enumeration. We believe therefore that the ability to group percepts into sets preceded enumeration because enumeration requires words to represent the numbers. Before our human ancestors acquired language they probably had the sense of quantity at a level shared by some non-human animals. But it was only with the acquisition of number words that they could actually count. Once again, this conclusion is strictly an abduction as we have no way of gathering evidence of how counting and the use of numbers began. We remind the reader of the experiment conducted by Brian Butterworth with aboriginal children that we reported in Chap. 1 that suggested "we have an innate system for recognizing and representing numerosities."

Hopefully the reader will allow us to make one more abduction, namely that the transitions from mimetic communication to verbal language and from percept-based thinking to concept-based thinking also represents the transition from pre-human hominids to fully human Homo sapiens. The Extended Mind hypothesis that we have just described to explain the origin of verbal language and its emergence from mimetic communication entails three distinct bifurcations:

1. the bifurcation from percepts to concepts,
2. the bifurcation from the brain to mind, and
3. the bifurcation from archaic Homo sapiens to full-fledged human beings, Homo sapiens capable of language and conceptual thinking (Logan 2007, 64).

The Emergence of Grammar: The Lexical Hypothesis

Although initial words of a spoken language were likely the strange attractors of the set of percepts associated with the concept represented by the word not all words arose in this fashion. Once a simple lexicon of words came into being there was a need to establish words that linked the words created from direct percepts. A new mental dynamic emerged in which abstract thoughts and concepts could be articulated and this required additional words to express the new relationships that would emerge with the new percept-based words. These new words would not have emerged as attractors of percepts but rather as representations of the abstract concepts of the grammatical relationships among words. The first words of this nature would have been, in all likelihood, associated with grammar and categorization and hence related to mathematics in mind. Examples of the former would be grammatical words such as: pronouns, adjectives, adverbs, conjunctions, pointing words such as this and that and examples of the words for sets of words or concepts such as words like: animals, people, birds, fish, insects, plants, and fruits.

Our hypothesis that human language began with the emergence of words acting as concepts representing sets of percepts falls within the tradition known as the lexical hypothesis that "*the lexicon is at the center of the language system* (Donald 1991, 250)." Language began with a simple lexicon without any grammatical structure and that the use over time of this simple lexicon gave rise to grammatical and syntactical structures. The grammatical and syntactical structures are also concepts. They are concepts that encompass relationships between words just as words are concepts that encompass relationships between percepts.

In a certain sense mathematics represents a grammar of numerical quantities. Mathematics began in terms of grouping items with similar properties and associating numerical quantities with these sets. Just as the semantics of spoken words led through grammaticalization to syntactical relationships one can imagine a similar process occurring within the confines of mathematical language.

Conclusion: The Mind as a Supervenient System

The mind is a supervenient system composed of

1. the brain,
2. the percepts of the brain and the nervous system,
3. verbal language consisting of words that act as strange attractors for those sets of percepts associated with each word,
4. the ability to think in terms of sets (i.e. the structures of set theory) and hence the beginning of mathematics in mind.

As with any supervenient system there is top down causality from the mind to its components and bottom up causality from the components of the mind to form the supervenient system of the mind in what is commonly called emergent dynamics. One cannot reduce the process by which the mind emerges to top down or bottom up causality as both forms of causality take place simultaneously. It is impossible to reduce a complex system to the behavior of its independent components.

Summing up our hypothesis of mathematics in the mind we would say that it is tied to the emergence of language and that the two co-emerged and then co-evolved. The mind which is language plus the brain has a mathematical structure by virtue of the fact that speech arose from the ability of Homo sapiens to think mathematically in terms of creating sets of percepts that were then represented by a word acting as a concept or a concept acting as a word.

The mathematical structure of the mind and the mind itself emerges simultaneously with language. By associating a set of percepts with a concept represented by a word the brain creates verbal language and bifurcates into the human mind. The brain, concepts, percepts, verbal language and a primitive form of set theory together form an emergent supervenient emergent phenomena, namely the mind. The mind = brain + language in the sense that the brain creates the structures of language out of the sets of percepts and language transforms the brain into a mind that is capable of creating words as sets of the brain's percepts. The supervenient system of the mind consists of the brain and the brain's percepts organized into sets as concepts, which are the words of the spoken language. There is the top down causality of the mind acting on the brain and its percepts to create its concepts and verbal language complete with its semantics and syntax. On the other hand, there is the bottom up causality of the brain, its percepts, and its mathematical cognitive structures organizing themselves into the human mind capable of language and the ability to conceptualize and deal with matters outside of the realm of the immediate here and now. As with all emergent dynamic phenomena it is impossible to sort out the top down and bottom up causality as they are happening simultaneously.

Chapter 3
Mathematics, Writing and Notation

Introduction

In Chap. 2 we argued that the first primitive forms of mathematical thinking in terms of a rudimentary set theory led to the emergence of verbal spoken language and that verbal spoken language allowed the formulations of numbers essential for counting and enumeration. In this chapter we will describe how enumeration and the accounting of agricultural commodities led to the emergence of writing and mathematical notation. It is only with the emergence of mathematical notation that mathematics blossomed. The mathematics of pre-literate hunting and gathering cultures was limited to simple counting and enumeration. All cultures that independently developed writing systems were agricultural societies.

We shall also explore in this chapter how writing and mathematics are languages in their own right that evolved from spoken language. We will demonstrate the way in which mathematical notation was essential for the evolution of mathematics. In particular, we will describe the invention of zero and the role it played in the development of the place number system, the concept of negative numbers, the notion of infinity as a mathematical concept, and the emergence of algebra. In later chapters we will argue that in addition to the three languages of speech, writing and mathematics that are the focus of this chapter that science, computing and the Internet can also be considered as languages that evolved from speech, writing and mathematics (Logan 2004b), but let us now turn to tell the story of the emergence of the languages of writing and mathematics.

R. K. Logan, I. Pruska-Oldenhof, *A Topology of Mind*, Mathematics in Mind,
https://doi.org/10.1007/978-3-030-96436-8_3

What Is a Language? A New Concept

A language is a system of communication and a medium for formulating thought. There are some 6000 languages spoken in the world today and that number keeps decreasing due to globalization. There were once many more languages before the technologies of the communication and travel shrunk the globe to the dimension of a village, a Global Village as McLuhan (1962) described it. But in addition to the variety of different spoken languages there are another class of languages, namely, writing, mathematics, science, computing and the Internet, the languages which we will show originated from the first language of speech and form an evolutionary chain of six languages (Logan 2004b).

Linguists have traditionally defined language in terms of communication. Spair (1921) for example defined language as "a purely human and non-instinctive method of communicating ideas, emotions, and desires by means of a system of voluntarily produced symbols." Although one of the main functions of language has been communication, this is not its only function. Language plays a key role in information processing, including its storage, retrieval, and organization. Language is a tool for developing new concepts and ideas; it is an open-ended system. Language is the medium in which thought takes place, at least abstract conceptual thought which is quite different than the non-abstract, non-conceptual thought of non-human animals none of whom possess verbal language. Since writing, mathematics, science, computing and the Internet permit the development of ideas that could never arise through the use of speech alone, we must consider these other modes as distinct languages, albeit related to the first language of speech.

We have therefore extended Sapir's definition of language as "a purely human and non-instinctive method of communicating ideas, emotions, and desires [to include the processing, storing, retrieving, and organizing information] by means of a system of voluntarily produced auditory and visual symbols." Speech is therefore not the only form of language. We suggest, for instance, that speech and writing are two distinct but related forms or modes of language. This differs from the beliefs of traditional linguists who consider that speech is the only form of language and that writing is merely a system for transcribing or recording speech. This point of view dates back to Aristotle: "The sounds… are symbols of ideas evoked in the soul and writing is a symbol of the sounds" (Bandle et al. 1958, 95). This sentiment was echoed by Ferdinand de Saussure (1967), one of the founders of the field of linguistics, who wrote: "Language and writing are two different systems of signs; the latter only exists for the purpose of representing the former… The subject matter of linguistics is not the connection between the written and spoken word, but only the latter, the spoken word is its subject." Leonard Bloomfield (1933, 21) held a similar opinion: "Writing is not language, but merely a way of recording language by means of visible marks."

Their definitions are restricted to a model in which the sole purpose of language is communication and do not take into account the information-processing capabilities of language, a key consideration for understanding the subsequent emergence of the languages of writing, mathematics, science, computing and the Internet. But a new generation of linguists who viewed language from the perspective of informatics have emerged that understand the multi-tasking nature of language.

Michael Stubbs has critiqued Bloomfield's definition of writing as follows: "Writing is not merely a record... I know from personal experience that formulating ideas in written language changes those ideas and produces new ones" (Stubbs 1982). Joyce Hertzler (1965, 444) concurs: "People often find that their thoughts are clarified and systematized, and that necessary qualifications and extensions appear, when they subject them to the more rigorous test of exactness and completeness demanded by the written form." Frank Smith (1982, 204) also agrees: "Ideas develop from interaction and dialogue... especially with one's own writing".

Written language is derived from spoken language, but because they process information so differently it is useful to regard them as two separate language modes. Arguments that support the notion of writing as a separate mode of language can also be made for mathematics, science, computing, and the Internet. These five additional modes of language each have unique strategies for communicating, storing, retrieving, organizing, and processing of information that differ from speech. We have therefore extended the notion that speech and writing as separate modes of language to claim that speech, writing, mathematics, science, computing, and the Internet form an evolutionary chain of languages which are distinct but interdependent. We shall return to this topic in the next two chapters but for now we can say that all of these languages including the Internet, the sixth language had their origin with speech, the first language. Each of the later modes of language are derived from and incorporate elements of the earlier modes of language that preceded them and from which they evolved.

Alphabet, Mother of Invention and the Alphabet Effect

There are lessons to be learned to help us understand the origin and emergence of speech by studying the evolution of notated language. McLuhan and Logan (1977) and Logan (2004a) showed that phonetic writing and in particular the phonetic alphabet played a key role in the development of deductive logic, abstract science, codified law, and monotheism. They showed that these five developments, which emerged between the Tigris-Euphrates Rivers and the Aegean Sea between 2000 and 500 BC, formed an autocatalytic set of ideas that supported each other's development. The alphabet not only served as a convenient way to notate speech it also taught the lessons of analysis (breaking up words into their basic phonemes), coding (writing or transcribing spoken language into a visual form), decoding (reading or transforming visual signs into verbal language) and classification (through alphabetization). This work revealed that language is both a medium of communication and an informatic tool since the structure of a language influences the way in which its users organize information and develop ideas.

The Six Languages of Speech, Writing, Math, Science, Computing and the Internet

Making use of what was learned from the alphabet effect, Logan (1995, 2004b) showed that speech, writing, mathematics, science, computing and the Internet represented an evolutionary chain of six independent languages.

These six systems of organizing information can each be considered as a unique language because each possesses its own unique semantics and syntax, which is how linguists define a language. Two linguists, Paivio and Begg (1981, 25) wrote, "semantics and syntax—meaning and grammatical patterning—are the indispensable core attributes of any human language".

The semantics or vocabulary of speech and writing are quite different. Written language contains many more abstract words than speech especially everyday speech. This is even true of a comparison of the vocabulary of the epic poetry of Homer that was part of the pre-literate Greek oral tradition and the written language of the ancient Greeks once they obtained alphabetic writing. Written Greek has many more abstract words than the orally recited epic poetry of Homer. It is also the case that the vocabulary of contemporary written language and spoken language is quite different. The same is true of grammar, the syntax of written language is quite formal and that of spoken language is more relaxed. In fact, if one were to transcribe an ordinary conversation one would find multiple grammatical errors. To convey ones meaning with writing one has to be precise as there are no other cues as to the meaning that is intended other than the written text. With spoken language there are many more cues from the tone of the speech and from the facial gestures, the hand signals, body language and the tone of the speaker all of which comprise what is called mimetic communication. A hint of the connection of grammar to written prose rather than spoken language is that the very word grammar is derived from the Greek word gramma (γραμμα) which means letters as in the letters of the alphabet. Given that alphabetic writing is a linear sequence of letters it is interesting that the word gramma (γραμμα) for letters is closely related to the Greek word for line, grammi (γραμμι). Grammar is a characteristic of the written word. A grammatical error in speech rarely gives rise to a misinterpretation as is sometimes the case with the written word because of the added information of tone, facial gestures, hand signals and body language that accompanies the spoken word.

The same argument that the uniqueness of syntax and semantics of the other four languages makes them independent languages on their own. The syntax of mathematics are mathematical operations and logic while the semantics are the numerals, various signs (such as +, −, x, =, <, and >) and mathematical operators such f(x) = sin x. In science the vocabulary might be borrowed from everyday speech but the word 'force' has a very limited meaning in science as that which cause an acceleration or a counter force whereas there are multiple meanings of the word force in everyday speech as in "I was forced to do that." Computers have their own language which we call computer language with its own unique operating systems that are quite different than that of spoken language. The Internet is still an altogether different language than speech or writing with its unique protocols ruled by the grammar of HTML and its own unique semantic elements such as hypertexted text.

The six forms of language, speech, writing, mathematics, science, computing and the Internet, form an evolutionary chain of languages because each new language emerged from the previous forms of language as a bifurcation to a new level of order in response to an information overload that the previous set of languages could not handle. Each new language built on the features of its predecessors while adding a number of new information-processing elements of its own.

The Origin of Spoken Language and Its Evolution into the Languages of Writing and Mathematics

Speech, the first form of human language, is the basis of all subsequent linguistic modes of communication and information processing. Spoken language is the sum of information uttered by human speakers. Written language, the sum of information which has been notated with visual signs, makes use of different grammatical constructions and vocabulary than spoken language. The language of mathematics also has its own unique vocabulary and grammatical structures. They differ from speech in that they involve permanent records, whereas speech disappears immediately after it is uttered. Writing and mathematical notations were the first forms of written language. Schmandt-Besserat (1978, 1992) showed that they both grew out of the system of book keeping in Sumer that was used to record the payments of agricultural tributes by farmers to priests dating back to 10,000 years ago when clay accounting tokens were used as receipts and were then recorded on clay tablets just over 5000 years ago.

We regard these two modes of information processing, writing and numerical notation as two separate languages, both of which are also distinct from spoken language. Speech, writing, and mathematics are distinct modes of language which form an evolutionary chain of development with writing and mathematics incorporating many elements of the spoken language that preceded it (Logan 2004b). What each of these three modes of language share is a distinct method for the communication and the processing of information. And each of them is deployed for a different function and purpose.

Before going into a detailed description of how writing and mathematical notation evolved from speech, we will first provide the reader with a timeline for the emergence of speech, writing and mathematics. Even before there was a spoken language there was the mimetic communication consisting of facial gestures, hand signals, body language and non-verbal sounds like grunts and sighs. This proto-language could only be used for communication as it was purely percept based and could only communicate about things in the immediate here and now. It was the mode of communication used by our pre-verbal hominid ancestors. It was followed by verbal spoken language some 50,000–100,000 years ago by the first fully human beings, i.e. homo sapiens, as was described in the last chapter.

Writing and mathematical notation emerged circa 3000 BC in Sumer and shortly thereafter in Egypt, India and China where it is believed the idea of writing was borrowed but not the specific form of the writing. Writing also emerged in the new world in Meso-America and Peru independently. The first forms of writing were logographic in which each word was represented by a unique visual sign. With time the writing system that originated in Sumer evolved into a phonetic syllabary of 60 signs in Babylon circa 1800 BC during the reign of King Hammurabi. Alphabetic writing emerged circa 1500 BC in the South Sinai among the Midianites or the Canaanites (Kenites in Hebrew Scriptures) which eventually led to abstract science and deductive logic in Ancient Greece circa 500 BC (McLuhan and Logan 1977; Logan 2004a).

The Origin of Writing and Mathematical Notation

We will argue that not only did mathematical thinking lead to verbal language but it also gave rise to written language through the development of mathematical notation. The very first notation for recording quantities were tally sticks in which the number of notches in the stick or antler corresponded to some quantity that the maker of the tally stick wanted to keep track of. The tally stick itself gave no indication of what was being tallied. Tally sticks, the origin of notation, can be traced back to 17,000 years ago to etched or notched sticks, bones or antlers (Marshack 1964). It is also believed that pebbles, grains, twigs, or shells were used as counters (Schmandt-Besserat 1984, 48–60).

Denise Schmandt-Besserat (1978, 1981, 1984, 1992) showed that the origins of writing and mathematical notation can be traced to the small three-dimensional clay tokens that were used for accounting purposes in the Fertile Crescent between the Tigris and Euphrates Rivers from 8000 to 3000 BC. The shape of each token determined what agriculture commodity was being enumerated. The tokens acted as receipts for tributes or taxes paid in agricultural commodities by farmers to priest who redistributed the farmer's produce to the irrigation workers whose labor made farming possible.

The token system remained basically unchanged for the first 5000 years of its use during which time farmers stored their receipts in the form of clay tokens in straw baskets. But then beginning sometime around 3200 BC in Sumer the tokens began to be stored in clay envelopes so the tokens would not be scattered and possibly lost. A hundred years later a very clever accountant-priest observed that every time they needed to see what tokens were inside the clay envelope they had to break it open and re-seal the tokens with a new clay envelope. He came up with the idea of pressing the clay tokens into the clay envelope before sealing it so that they would know what was inside. It was not long after this system was in use that the accountant-priests realized that they did not need to put the tokens into the clay envelopes because they knew what was inside by reading the envelope and voila the clay tablet was born.

As urbanization took hold in Sumer and larger amounts of commodities had to be recorded it became tedious to press a clay token into the wet clay envelope multiple times so a new system of accounting emerged. The two logograms for the small and large measure of grain, the ban and the bariga, were used to represent the numbers 1 and 10 respectively. Let us say a certain number of jars of oil were to be recorded on the envelope. That number was represented by the appropriate combination of bans and barigas and placed beside the sign for a jar of oil. The ban and bariga operated like I and X in Roman numerals. In order to determine which shapes of the token represented the numbers and which the commodity being enumerated the accountant-priests continue to press the ban and bariga tokens into the clay and the commodities being enumerated were represented by etching in the clay with a stylus the shape their token would have made when pushed in to the clay envelope. In this way a separation was made for signs representing words, the name of the commodities being enumerated, and the numbers representing the number of those particular commodities. This was the very beginning of writing and mathematical notation on the surface of the clay envelope.

It was soon recognized that if the information of what was inside the envelope was on its surface it was not necessary to actually put the tokens into the envelope. It was at this point that the rounded clay envelope morphed into a flat two-dimensional tablet with the impressed and etched contours of tokens upon their upper surface.

It is important to note that the impressed and etched logograms, semantically, were no different from the three-dimensional tokens from which they emerged. The logograms were two-dimensional negative imprints of the three-dimensional tokens. They were immediately recognizable to the users of the token system because they shared the same outline and bore the same markings as the actual tokens. In fact, the first tablets are not true writing but merely the permanent records of an accounting system based on tokens in which the tokens themselves were actually discarded or used again on another tablet once their imprints had been made. From the point of view of function and informational content, impressed logograms on clay tablets operated exactly the same way as the three-dimensional token system. The medium changed, however, and the way that information was displayed changed radically from a collection of palpable three-dimensional clay objects to abstract symbols arrayed in two-dimensional patterns on the flat surface of a clay tablet. The shift from three-dimensional artifacts to logograms in the form of negative images impressed on two-dimensional surfaces triggered a chain reaction that resulted in the invention of logographic writing, phonetic coding, and abstract numerals.

It was not very long after the use of tablets for accounting that it was realized that not only the names of agricultural commodities could be represented by visual signs but that any word in their spoken language could also be written as an etched sign on a clay tablet. One of the reasons for relating this story is that we learn that writing and math notation emerged simultaneously on the clay tablets of accountants some 5000 years ago. One of the conclusion we can draw from this is that if one can read well one should be able to do math and vice-versa if one can do math one should be able to read and write well. The lesson we would draw from this is that instruction in primary schools of reading and writing and mathematics should be more integrated.

Numerical notation and writing systems have always been closely linked. In fact, the first forms of notation were numerical and took the form of tallies etched on bones as we noted above. This might explain the fact that in the Semitic languages the words for "scribe" and "count" are the same, i.e., SPR.

Writing and the notation for abstract numbers emerged in Sumer at precisely the same moment in history, from a common progenitor, clay accounting tokens. The use of these tokens created a new skill set that led to the further evolution of language.

> The manufacture and manipulation of tokens and impressed signs promoted the development of hand-eye coordination and fine motor skills essential for writing. More importantly, tallies, and especially tokens, introduced the visual bias that characterizes written notation. Tokens stressed the importance of uniformity. They were the first class of artifacts whose uniformity of shape was key to their function. Tokens were used as repeatable symbols in much the same way that numerals and the letters of the alphabet are. The visual bias created by the use of uniform and repeatable tokens increased when the tokens were used to imprint the outer surfaces of clay envelopes and, later, the impressed tablets. The pattern of token markings arranged in neatly spaced lines within a two-dimensional array permitted the user (or reader) to encompass the whole field of data in a single glance.
>
> Tallies, tokens, and impressed signs also paved the way to writing and abstract numerals on the cognitive level. Each of these methods of record keeping can be regarded as phases in the development of abstract notation. Impressed token markings increased the level of abstraction. The two-dimensional format of the impressed tablets was more abstract than the three-dimensional tokens in that the negative impression of the tokens could not be grasped in the hand and manipulated like three-dimensional tokens. The loss of volume, concreteness, and tactility removed the symbols one step further from the physical three-dimensional reality for which they stood. The display of information within a two-dimensional array, however, permitted the user to see the information in a new light. By being able to observe more data at a single glance, the user of the tablet system began to think more globally about the data. More abstract patterns of classifying and analyzing the data became possible. If the medium is the message, then the message of the two-dimensional display of information on tablets was that of abstraction, classification, analysis, uniformity, repeatability, and the power of the visual. (Logan 2004b, 90–91)

With a written notation for both words and mathematical notation not only was communication enhanced but mathematical thinking became more sophisticated. De Cruz and De Smedt (2013) argue that

> mathematical symbols are not only used to express mathematical concepts—they are constitutive of the mathematical concepts themselves. Mathematical symbols are epistemic actions, because they enable us to represent concepts that are literally unthinkable with our bare brains. [Signalling] an intimate relationship between mathematical symbols and mathematical cognition.

Thus, mathematical thinking gave rise to mathematical notation and writing which in turn led to the further development of mathematical thinking. Here we encounter a spiral structure to human cognition which is a theme we will visit many times in our study.

The Origin of Schools of Instruction for Reading, Writing and Arithmetic

School days, school days, Dear old golden rule days.
'Readin' and 'ritin' and 'rithmetic, Taught to the tune of a hick'ry stick.

Education before the emergence of writing and mathematical notation was strictly apprenticeship learning on the job by watching a master at work and assisting the master and slowly learning and acquiring the skills of the master. There were no schools so to speak. But with reading, writing and mathematics schools were organized to teach these skills. The first schools were organized in Sumer with the purpose of training scribes. The second system of formal education to emerge was that of the ancient Egyptians who were the second culture to create a writing system complete with arithmetic notation. The same pattern emerged wherever literacy and numeracy took hold in ancient Greece, India and China.

Literacy and numeracy required formal education. One could not learn these skills by watching someone else perform these tasks and then imitate them. The secret codes of the letters and the numerals and their proper use had to be taught and that required teachers who became scholars and scientists as they collected information for their lesson plans in the form of lists of natural phenomena such as trees, insects, rivers, and minerals; lists of geopolitical features such as cities and rulers; and mathematical tables all of which were used as teaching tools. The first scientists were the teachers who compiled these lists. The vanguard of Sumerian science were the teachers in the scribal schools.

> Within its [the school's] walls flourished the scholar-scientist, the man who studied whatever theological, zoological, mineralogical, geographical, mathematical, grammatical, and linguistic knowledge was current in his day, and who in some cases added to the knowledge (Kramer 1959, 2).

The Fourth Language of Science

As we just learned the second and third languages of writing and mathematics emerged from the first language of speech. These two forms of notation were invented by accountant-priests to deal with the information overload that speech alone could not handle. The administrators of the agricultural system could not keep track of all the transactions with the farmers paying their taxes so they first created a system of using three-dimensional clay tokens as a way of extending their memories. The use of the clay tokens over time led to writing and mathematical notation, which in turn led to a school system to teach the new skills of literacy and numeracy. The teacher/scholars then created new forms of knowledge that were organized in a systematic way making use of writing and mathematics, which gave rise to the fourth language of science.

The next step in the emergence of written language after the logographic writing developed in Sumer was the emergence of phonetic writing in Babylon in the form of a phonetic syllabary of 60 signs. The original writing system that appeared in Sumer was used to transcribe a non-Semitic language. When the Akkadians, a Semitic people conquered the Sumerians they took over their writing system and applied it to their Semitic language. For a while the Akkadian culture was bi-lingual and as a result some of the signs in the Sumerian written language were used phonetically to represent Semitic words. Over time the writing system introduced more and more phonetic elements so that by the time of the reign of Hammurabi in Babylon circa 1800 BC, a reform of the writing system took place in which all the words of the language were transcribed in terms of a 60-sign phonetic syllabary. It was at this time that Hammurabi introduced standardized weights and measures and the Hammurabic code, a legal code that became a model for law as found in the Torah (the first five books of the Bible) and in the legal code of ancient Greece, Rome and subsequently all of the legal codes of Western culture.

The goals of the science that developed in Babylon and in Egypt by scholar teachers were quite modest and consisted of the gathering and organization of practical knowledge. No serious attempts were made to engage in abstract science as was practiced by the ancient Greek beginning circa 500 BC. That development as we shall see in Chap. 4 was stimulated by alphabetic writing.

Phonetic Writing and the Alphabet Effect

The Egyptian whose hieroglyphic writing system was logographic also introduced a 22 letter set of uni-consonantal signs that were strictly phonetic and were used in a very limited way strictly for rendering foreign proper names, some prepositions, grammatical suffixes of logographic words and a very small number of frequently used words. These phonetic signs were also logographic representing some object but the sign was not used to represent the object itself but rather the sound of the first phoneme of the spoken word associated with the object. If one were to create a similar system using the English language a logogram of a ball would represent the sound of b and a logogram of a dog would represent the sound of d and so on and so forth.

The Midianites who lived in the Sinai Desert were copper miners and smiths who traded with the Egyptians. They are the Canaanite people identified in the Hebrew scriptures as the Kenites whose name means coppersmiths. Their high priest was Jethro with whom Moses lived after he was exiled from Egypt. The Kenites asked the Egyptians to teach them their writing system but the Egyptians presumably to keep their sacred writing (the literal translation of hieroglyph) secret only taught them about their 22 uni-consonantal sign system, which the Midianites adapted for their language. This story is consistent with the discovery of the oldest known phonetic alphabetic inscription that were discovered by Petrie (1906) at the Serabit el Khadem temple in the Sinai. Other inscriptions have been found in or near the copper mines in the Sinai and are thought to have been left by the Kenites, who worked these copper mines.

The proto-Sinaitic alphabet spread quickly to other nations through trading relations. The first cultures to adopt it as it were the Semitic cultures, most significantly, the Israelites or Hebrews and the Phoenicians, and much later by the speakers of Aramaean and Arabic who were able to used it in basically the same form as the proto-Sinaitic alphabet. The Greeks, who traded with the Phoenicians adopted the Phoenician alphabet and adapted it to the needs of their language.

The first two letters of the Proto-Sinaitic alphabet were alef and beit, which are the Semitic words for ox and house respectively. They correspond to the first two letters of the Greek alphabet alpha and beta written as α and β. The word alphabet is derived from these first two letters of the Greek alphabet, alpha and beta. The diagram below represents the first two letter of the Proto-Sinaitic representing the head of an ox and a house. The Greek letter alpha, α, has the two horns of the ox facing side-way and the upper case letter A has the two horns facing downwards. The Greek letter beta, β, and the upper case form B, each look like a house as is the case with the Midianite letter beit, but they have two rooms.

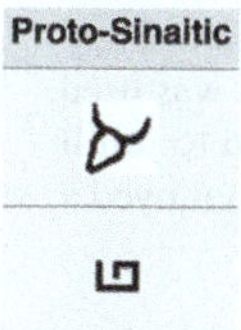

The reason for the correspondence of the Greek alphabet and the Midianite one is that the Greeks adopted the Phoenician alphabet which is basically the same as the Midianite alphabet. The Greeks had to add some additional letters to represent their vowels. In Semitic languages the three consonants of which a word is composed completely and uniquely specify that word. The internal vowels are used to make grammatical distinctions as is the case with the English words sang, sing, song and sung. Indo-European words require vowels to be transcribed. Imagine how we would be able to distinguish the words idea, do, due, and die without vowels. All four words would have to be written as d if there was no way to represent the vowels a, e, i, o and u. The aspirant alef was transformed into the vowel a and the letter hay into the vowel e. The Greeks also had to add certain letters like phi, φ, and psi, ψ, as these sounds do not exist in the Semitic languages. But other than these additions and transformations the letters of the proto-Sinaitic alphabet were used in the Greek alphabet including the order in which the letters appear. The Greek alphabet then became the model for all the languages of Europe. The Phoenician alphabet through

Aramaean played a similar role for the languages of the Middle East and South Asia including Persian, Sanskrit and the other languages of the Indian sub-continent.

Alphabet and Monotheism

The impact on the Hebrews of the alphabet which they borrowed from the Midianites was immediate and dramatic. In addition to bringing writing to the Hebrews, Moses also brought them codified law, or the Ten Commandments, as well as a more abstract and monotheistic concept of God. These three developments occurred at the same time, as recounted in the Bible by the story of Moses receiving the Ten Commandments on Mount Sinai: "He gave Moses, when He had made an end to communing with him upon Mount Sinai, two tables of testimony, two tables of stone, written with the finger of God" (Ex. 31:18).

> The appearance of the phonetic alphabet among the Semitic peoples in Canaan and Sinai in the middle of the second millennium B.C. marked the beginning of a new period of creativity that would take Western civilization beyond the Sumer-Babylonian cultural achievements of syllabic writing and codified law. The Proto-Canaanite (or Proto-Sinaitic) alphabet based on twenty-two uni-consonantal signs was adopted by a number of Semitic nations, including the Canaanites, the Phoenicians, the Midianites, the Aramaeans, the Nabateans, and the Israelites or Hebrews. Of all the peoples whose writing systems directly descended from the Proto-Canaanite alphabet, the Hebrews were most influenced by the Alphabet Effect. They became the People of the Book, who expressed their culture almost exclusively through the written word and, hence, the alphabet… Writing among the Hebrews was used for both commercial and literary purposes. All Israelites were encouraged to read to fulfill their religious duties so that by the sixth or seventh century B.C., Hebrew society enjoyed a basic level of literacy (Logan 2004b, 92).

The main contribution of the Hebrews was monotheism, which we would contend made a contribution to scientific thought especially of the Greeks. The idea that the forces of nature were part of a unified system created by God rather than the contending forces of the many gods of the Pantheon, each in charge of a different domain of the world perhaps influenced the Greek philosopher/scientists who described the world in terms of a ruling principle whether that was the Water of Thales, the Apieron of Anaximander, the Air of Anaximenes, the Fire of Heraclitus or the Numbers of Pythagoras.

The Hebrews were also the first people to create a non-mythological history of themselves. All other histories—those of the Babylonians, the Egyptians, the Greeks—trace the origins of their people to the beginning of the universe and they glorify their leaders. The story of the Hebrews as recorded in the Bible are more realistic. They did not glorify their leaders as was the case with their neighbors. Noah was a drunkard. Moses smashed the tablets given to him by God and was punished by not being permitted to the Promised Land. King David lusted after another man's wife. The Hebrews therefore set standards for honesty and pursuing the truth, an essential element for both abstract science and deductive logic.

The Role of the Alphabet in the Invention of Zero and the Place Number System

The effect of the phonetic alphabet on the Greeks and Hindu mathematicians was even more significant for our understanding of mathematics in the mind than it was for the Hebrews, the Phoenicians and the Midianites. First, we turn to the Hindu mathematicians and show how their use of the alphabet helped them to discover zero and the place number system. In the next chapter we will recount how the phonetic alphabet was key to the Greeks development of deductive logic and abstract science.

The ordering of the letters of the alphabet had a direct impact on the development of the notion of zero by the Hindu mathematicians circa 500 AD and hence on mathematics in general. Before developing the idea of zero, Hindu mathematicians used the letters of the alphabet to represent their numerals. The first ten letters of the alphabet represented the numerals 1–10 and the next eight letters represented 20, 30, 40–90. Two more letters were used to represent 100 and 1000.

Once the idea of zero was introduced Hindu mathematicians created the place number system that used only ten numerals, the first nine letters of their alphabet to designate the numerals 1–9 and the symbol for zero which they called sunya, which literally means "leave a space". The way they arrived at the system was to consider the results of an abacus calculation. They used the zero sign whenever one of the columns of the abacus was empty. The way an abacus works is that the columns represent ones, tens, hundreds, thousands and so on and so forth. Let us say they arrived at a result where the hundreds column had three beads, the tens column no beads and the ones column had two beads indicating the result of 302. If they used the old numeral system, they would have written 302 as follows: the sign for 3 (the 3rd letter of their alphabet) followed by the sign for 100 (the 11th letter of their alphabet) followed by the sign for 2 (the 2nd letter of their alphabet). With the new place number system using the zero sign or the sunya sign they wrote 302 as 3 'sunya' 2 where sunya means literally "leave a space" indicating three hundreds + no tens + two ones. At first, they notated the sunya sign with a dot so 302 was written as 3.2. Over time the dot grew into the circle, 0 and 3.2 became 302. The place number system, which greatly simplified arithmetic operations like addition, multiplication and division. With the nine numerals from 1 to 9 and the sunya or zero sign all number from 0 to infinity could be represented by some combination of the ten numerals. The system also led to negative numbers. They represented negative 7 or – 7 by placing a sunya sign over the numeral 7 so that it appeared as 7 below 0: $_{7}{}^{\cdot}$ and then $_{7}{}^{0}$.

The key to the Hindu mathematicians developing the notion of zero was that the order of their alphabet which was fixed allowed them to associate the first nine letters of the alphabet with the numeral for 1–9. Making use of Roman numerals would have made the development of zero more difficult if not impossible.

It is worth mentioning that the Hindu place number system, like the idea of leaving a space when there were no tens in a number like 302 had been used earlier in

Babylonian sexagesimal arithmetic based on the number 60. The Babylonian system was a place number system in the sense that they notated numbers side by side representing multiples of $3600 = 60^2$, 60 and 1 respectively. If a number was x times 3600 plus y times 1 they would notate it as x \\ y where \\ indicated there was no multiple of 60 (Kaplan 2000, 4–13). They only used \\ between two numerals, never at the end so there was an ambiguity if xy was x times 60 + y or x times 3600 + y times 60. The Babylonian did not have an alphabet so their use of \\ never developed into the Hindu idea of sunya.

Let us demonstrate how the place number system facilitated calculations. Let us compare some simple calculations using place numbers with the same calculations using the Roman numerals consisting of the signs I, V, X, L, C, D and M representing 1, 5, 10, 50, 100, 500, and 1000 and the convention that IV is 4; VI is 6; XL is 40; and LX is 60.

Consider the product 56 × 6 = 336 using Hindu-Arabic versus Roman numerals.

$$\begin{array}{r} 56 \\ \underline{\times 6} \\ 336 \end{array}$$

versus the same calculation using Roman numerals.

$$\begin{aligned} \mathrm{LVI} \times \mathrm{VI} &= \mathrm{L} \times \mathrm{V} + \mathrm{V} \times \mathrm{V} + \mathrm{I} \times \mathrm{V} + \mathrm{L} \times \mathrm{I} + \mathrm{V} \times \mathrm{I} + \mathrm{I} \times \mathrm{I} \\ &= (\mathrm{L}+\mathrm{L}+\mathrm{L}+\mathrm{L}+\mathrm{L}) + (\mathrm{V}+\mathrm{V}+\mathrm{V}+\mathrm{V}+\mathrm{V}) + \mathrm{V} + \mathrm{L} + \mathrm{V} + \mathrm{I} \\ &= \mathrm{CCCXXXVI} \end{aligned}$$

where L × V = L + L + L + L + L; V × V = V + V + V + V + V; I × V = V and I × I = I; L + L = C and V + V = X.

The calculations involving long division or fractions become even more complicated with Roman numerals. The advantages with Hindu-Arabic numbers are obvious.

Arabic numerals are obviously the most abstract numerical notation possible just as the alphabet is the most abstract form of writing. It is ironic that the alphabet achieves its abstraction through phonetization whereas the Arabic numerals are logograms or ideograms that represent ten numerical values including zero. The letters of the alphabet and the Arabic numerals, however, share four important features that enable them to act abstractly:

1. Each system contains a small number of elements; 26 letters (for the English alphabet) and 10 numerals.
2. Both systems form a complete set. The total set of possible spoken words can be represented alphabetically by its 26 signs and any number, no matter how large it may be, can be represented in terms of some combination of the ten numerals 0–9.

3. The individual elements of the two systems, the letters and the numerals, are atomic. That is, they are identical and repeatable.
4. The values (sound or numerical) of the aggregate elements (the words or numbers) of the system depend not only on the atomistic components (the letters and numerals) of which they are made up but also on their order. In other words, both the letters and their order determine a word and both the numerals and their order determine a number. For example, ON is not the same as NO nor is 18 the same as 81.

Chapter 4
Mathematics, Deductive Logic and Abstract Science

Alphabet Mother of Invention

The phonetic alphabet had a profound effect of the development of abstract science and deductive logic in ancient Greece. McLuhan (1964) applying his notion that the medium is the message showed that the alphabet is much more than an economical transcription of speech into a written code. The alphabet also has subliminal effects on its users in that it provides a model for analysis, abstraction, coding, and decoding that characterize scientific thought. Of all the writing systems, the phonetic alphabet permits the most economical transcription of speech into a written code. It also introduces a double level of abstraction in writing as words are decomposed into their meaningless phonemes, which are then represented by a small number (22–26) of the letters of the alphabet. The letters are visual signs that are as meaningless as are the phonemes they represent. The alphabet therefore encourages abstraction and analysis as each word must be broken down into its basic phonemes. It also encourages coding when words are represented by the meaningless signs or the letters of the alphabet and decoding when an alphabetic text is read.

McLuhan and Logan (1977) in an article entitled "Alphabet, Mother of Invention," indicated some of the cultural impacts that resulted from the invention of phonetic writing. They suggested that:

> Western alphabetic and Chinese literacy represent the two extremes of writing. The alphabet is used phonetically to visually represent the sound of a word. Chinese characters are used pictographically to represent the idea of a word. Consequently, they are less abstract and less specialized than alphabetic writing. Eastern and Western thought patterns are as polarized as their respective writing systems.
>
> Western thought patterns are highly abstract, compared with Eastern. There developed in the West, and only in the West, a group of innovations that constitute the basis of Western thought. These include in addition to the alphabet: codified law, monotheism, abstract science, formal logic, and individualism. All of these innovations, including the alphabet, arose within the very narrow geographic zone between the Tigris-Euphrates river system and the Aegean Sea, and within the very narrow time frame between 2000 B.C. and 500 B.C. We do not consider this to be an accident. While not suggesting a direct causal

R. K. Logan, I. Pruska-Oldenhof, *A Topology of Mind*, Mathematics in Mind,
https://doi.org/10.1007/978-3-030-96436-8_4

connection between the alphabet and the other innovations, we would claim, however, that the phonetic alphabet played a particularly dynamic role within this constellation of events and provided the *ground* or framework for the mutual development of these innovations.

The effects of the alphabet and the abstract, logical, systematic thought that it encouraged explains why science began in the West and not the East, despite the much greater technological sophistication of the Chinese—the inventors of metallurgy, irrigation systems, animal harnesses, paper, ink, printing, movable type, gunpowder, rockets, porcelain, and silk.

Credit must also be given to monotheism and codified law for the role they played in developing the notion of universal law, an essential building-block of science. Almost all of the early scientists—Thales, Anaximenes, Anaximander, Anaxagoras and Heraclitus—were both law-makers in their community and monotheistically inclined. They each believed that a unifying principle ruled the universe.

The Role of the Alphabet in the Emergence of Deductive Logic by the Ancient Greeks

One of the crowing achievements of Greek thought that emerged shortly after they adopted the phonetic alphabet was abstract science and deductive logic. The phonetic alphabet promotes abstract thought more so than a logographic writing system like that used by the Sumerians and the Egyptians. The words of the written language are composed of meaningless signs so that there is no connection between word and the letters of the alphabet that are combined to create the word. Alphabetic writing also promoted analysis because each time one hears the sound of a word that one wishes to transcribe one must analyze the word into the basic phonemes that make up the word and then represent those phonemes with meaningless visual signs. Writing with an alphabet involves coding and reading involves decoding.

If one compares the vocabulary that emerged with alphabetic literacy with the vocabulary of the Greek oral tradition of the epic poetry as exemplified by Homer's Iliad and The Odyssey, one finds a plethora of abstract notions in alphabetic based literature. Examples of words found in alphabetic literature and not found in Homer that are still in use to this day include body, atom, matter, essence, space, translation, time, motion, permanence, change, flux, quality, quantity, combination, ratio, psyche, logic, physics (from phusis, Greek for nature), philosophy (literally love of wisdom), analysis, epistemology, metaphysics, ontology, axiom, theorem, syllogism, line, point, angle, add, subtract, multiply, divide and ratio. These terms and concepts became the language of philosophy, science and mathematics. A rational approach to analyzing problems logically and finding solutions to them developed with the alphabet. Ideas such as truth, beauty, justice, and reason took on new meanings and became the subject of a new type of discourse. The alphabet also provides a natural classification scheme as all the words in one's language can be ordered alphabetically.

In addition to serving as a paradigm of abstraction and classification, the alphabet serves as a model for division and fragmentation. Deductive logic entails breaking an argument up into its distinct elements consisting of two or more axioms and an inferred conclusion. An example is the following syllogism: Socrates is a man, all men are mortal, therefore Socrates is mortal. With the alphabet every word is fragmented into its constituent sounds and constituent letters. The Greeks' idea of atomicity, that all matter can be divided up into individual distinct tiny atoms, is related to their use of the alphabet: "Atomism and the alphabet alike were theoretical constructs, manifestations of a capacity for abstract analysis, an ability to translate objects of perception into mental entities (Havelock 1976, 44)."

With alphabetic writing knowledge begins to take on an identity separate from the knower. "The phonetic alphabet was a technical means of severing the spoken word from its aspect of sound and gesture... The literate man or society develops the tremendous power of acting in any matter with considerable detachment from the feelings or emotional involvement that a non-literate man or society could experience" (McLuhan 1964, 193 and 79). The Greeks through writing developed a notion of objectivity—the separation of the knower from the object of his study. This is the beginning of the scientific method and the source of the dichotomy the Greeks created between subjective and objective thinking. Poetry and art are examples of subjective thinking, whereas philosophy and science are examples of objective thinking.

The abstract and systematic nature of the alphabet provided an environment, a way of thinking, a mind-set that encouraged deductive logic and abstract science. The matching of the sound of a word with the meaningless letters of the alphabet encouraged the matching of an axiom and an inference with a conclusion. The linking together of the elements of the alphabet, the letters, to form words provided a model for the linking together of ideas to form a logical argument. Arguments are linked together in order to reach a conclusion. The deductive reasoning implicit in formal logic and geometry also formed the basis of early Greek science, which attempted to derive its description of nature from first principles.

The Egyptian that employed a logographic writing system were content to create lists of geometric relationships that they determined empirically were true. The Greek, on the other hand, had to prove that these relationships were true based on what they considered to be self-evident axioms. They engaged in a matching exercise with their axioms that to them were self-evidently true and their theorems.

The use of the alphabet serves as a model for matching, an activity crucial for the development of logic. Each letter of the alphabet is matched with a sound and vice versa. Every time a word is read, a match between a visual sign and a spoken sound is made. Matching forms is the basis of rationality or logic. Rationality grew out of the concept of ratios, which in turn involves matching. A is to B as C is to D is an example of a ratio achieved by matching.

The discovery of zero is somewhat of a mystery. The ancient Greeks who developed deductive logic and axiomatized geometry never developed the concept of zero. The honor belongs to the Hindu mathematicians who invented zero more than 2000 years ago. Their discovery led them to positional numbers, simpler arithmetic calculations, negative numbers, algebra with a symbolic notation, as well as the notions of infinitesimals, infinity, fractions, and irrational numbers as described in the last chapter. The Greeks rejected all of these mathematical notions. They argued that one was a unity and therefore could not be divided. Their adherence to logic got in the way of the development of their mathematical ideas. The Hindu mathematicians were not so rigorous as far as logic went but they were far more imaginative and hence creative.

The Intimate Connection of Geometry and Logic

We most often associate geometry with the Greeks and especially Euclid's Elements of Geometry. Euclid, often referred to as the Father of Geometry, deduced the theorems of geometry from a small set of axioms circa 300 BC long after the mathematics of geometry had been developed by Greek mathematicians dating back at to Thales circa 600 BC. The results or theorems of axiomatic geometry of Euclid were actually arrived at empirically by the Egyptians circa 1650 BC and before the Egyptians, by the Sumerians and the Babylonians, whose geometric thinking has been dated back to 3000 BC (http://it.stlawu.edu/~dmelvill/mesomath/3Mill/chronology.html, accessed June 2, 2017).

Geometry in Egypt arose out of the need to measure the area of land in the possession of a landowner before the yearly inundation of the Nile. Rather than restore the boundary lines between properties that were destroyed by the flooding, each landowner was provided with a new plot of land more or less in the same location as before and with a total area exactly equal to the amount of land in his possession before the flood. Any extra land that was created by the flood belonged to the Pharaoh. Because of this need to measure the area of land accurately, an empirical science arose in Egypt called geometry, which literally means earth (geo) measuring (metry). Egyptian geometry is not derived from a set of axioms. There are no theorems or proofs or propositions. There are merely a set of rules that are used strictly for practical applications such as land and volume measurements and also needed for construction projects.

The Egyptians needed to know how much grain could be stored in a cylindrical silo so they also derived rules for measuring the volume of a cylinder based on the diameter of the cylinder and its height. They used an approximation for pi (π), namely the ratio of 256 to 81 which they arrived at empirically. The fraction 256/81 has an approximate value of 3.1605, which is quite close to the actual value of pi which is approximately 3.1459. The Egyptian value for pi had an accuracy of less than two-thirds of a percent or 0.63%. The Babylonians actually had a slightly more accurate value of pi equal to 25/8 = 3.125 with an accuracy of 0.53%. The Greek mathematician and scientist Archimedes sometime in the second century BC, surpassed the accuracy of both these estimates with the value of the ratio of 211,875 to 67,441 = 3.14163 with an error of just over 0.01%. The reader should note that the values for pi were expressed as the ratios of very large numbers because neither the Babylonian nor the Egyptians had the decimal system or place numbers because zero had not yet been invented by the Hindu mathematicians. That development did not occur several centuries after the flourishing of Greek geometry in approximately 500 AD and did not arrive in Europe until the thirteenth century.

The Egyptian sets of mathematical rules, as found in the Rhind papyrus for example, were used strictly for practical applications such as land measurements, volume measurements or construction calculations. There was no attempt to derive their formulas from first principals. They were strictly empirical rules that worked at an acceptable level of accuracy. The Rhind papyrus has been dated back to 1650 BC and is attributed to a scribe named Ahmose (Schneider 2006).

The Greeks, on the other hand, almost from the instant they became involved with geometry, were interested in systematizing and formalizing the Egyptian results. Greek geometry began with Thales, one of the seven wise men of antiquity, who was familiar with the Egyptian formulas as contained in the Rhind papyrus and is said to have travelled to Egypt where he became acquainted with Egyptian mathematics. He was the first to devise formal proofs of Egyptian geometric results. "But whereas in Egypt mathematics like ethics and medicine had been developed empirically and stopped short of philosophy, it became to Thales a means of discarding allegory and myth and advancing universal generalization" (Innis 1972, 64).

There are four theorems attributed to Thales by those ancient Greek philosophers who read his manuscripts that have since been lost and have described what he wrote. The four theorems are as follows:

- The diameter of a circle exactly bisects the circle
- The base angles of an isosceles triangle are equal
- The two pairs of angles formed by two intersecting straight lines are identical
- If one side and the two adjacent angles of a triangle are shared by another triangle, the triangles are identical.

Thales is credited with measuring the height of the pyramid by using the results of his theorems plus observing the length of the pyramid's shadow when the length of his own shadow was equal to his own height.

What the Greeks did beginning with Thales and culminating with Euclid was to show that the Egyptian empirical formulas of geometry like the Pythagorean theorem could be derived from a small set of axioms that they considered were self-evidently true. An example of an axiom used by the Greeks was the notion that the shortest distance between two points was a straight line. Actually this so-called obvious truth actually holds only for the distance between two points on a plane surface. The Greek mathematicians by formalizing Egyptian geometry by proving a number of empirically derived theorems from a set of first principles or axioms extended and generalized the Egyptian results. Thales method of proving theorems using basic axioms led to the Greek's development of deductive logic. "By discovering mathematical theorems, the Greeks came across the art of deductive reasoning. In order to build their mathematical knowledge they came to conclusions by reasoning deductively from what appeared to be self-evident (http://www.ancient.eu/Greek_Science/, accessed March 5, 2017)."

Greek mathematicians not only developed axiomatic geometry but they also invented deductive reasoning which gave rise to another branch of mathematics, namely deductive logic. Deductive logic was first used by Parmenides. It was eventually formalized by Aristotle in what he defined as syllogistic or term logic. A somewhat related form of logic known as propositional logic was developed by the Stoics beginning with Zeno of Citium and culminating in the work of Chrysippus.

For Aristotle a syllogism is an inference in which a proposition known as the conclusion follows by necessity from two other propositions that are premises and are assumed to be obviously true. For example, the two premises, 'Socrates is a man' and 'all men are mortal' leads by necessity to the conclusion that 'Socrates is mortal'.

Propositional logic or propositional calculus entails the study of propositions that can be either true or false and that are the products of other propositions to which they are connected to logically. The truth value of the proposition is then determined by the truth value of the component propositions of which they are composed.

Aristotle's syllogistic logic held swayed for more than two millennium, until the development of modern predicate logic beginning with Gottlob Frege who in 1879 published his *Begriffsschrift, eine der arithmetischen nachgebildete Formelsprache des reinen Denkens* (*Concept-Script: A Formal Language for Pure Thought Modeled on that of Arithmetic*).

The Role of Logic in the Formulation of Abstract Science

Logic as Marshall McLuhan (1987, 404) long ago pointed out has both service and disservice. He wrote "that any product or innovation creates both service and disservice environments which reshape human attitudes." The service of syllogistic logic is hard to dispute. It has provided service since its inception over 2500 years ago but it also produced a bias in the thinking of the ancient Greeks that had a negative effect on their scientific thinking and their development of mathematics.

Using logic and arguing that non-being could not be they sabotaged any attempt the Greeks would have made to formulate the notion of zero. The Hindu mathematicians, on the other hand, who were not so logically rigorous, were able to formulate the concept of zero and hence place numbers.

The strict adherence to logic also impacted Greek science. Before Parmenides came on the scene the early pre-Socratic philosopher/scientists valued the role of observation for gathering knowledge of the world. Heraclitus wrote, "The things of which there can be sight, hearing and learning—these are what I especially prize". "Eyes are more accurate witnesses than ears". Examples of the empirical approach of many of the pre-Socratic philosophers include an expedition Anaxagoras made to observe the remains of a shooting star, which he found was basically a rock with bits of metal embedded in it.

The service to science that logic inspired by axiomatic geometry contributed was to provide a model for causality. If axioms lead to theorems using deductive logic, then action or effects are derived from causes. The chain of logic became the chain of causality from causes to effects and the idea of a "first mover" and an organizing principle upon which the universe emerged. The pre-Socratic philosophers were basically materialist who attempted to explain all natural phenomena in terms of an organizing first principle.

Thales, for example, believed all things to be derived from water as water seemed to be a component of all forms of matter. For Anaximander, the first principle was the neutral substance Apieron from which opposites emerged. Anaximenes described all of nature in terms of air that could be found in nature with different densities, not unlike our idea that all matter is composed of proton, neutrons and electrons of different densities and configurations. Heraclitus used fire as his first principle, which is not all that different that our notion that with the equivalence of mass and energy ($E = mc^2$) that the universe is composed of energy, which Heraclitus expressed with the metaphor of fire. The atomist, Leucippus and Democritus, described nature in terms of indivisible atoms and the void. Empedocles held that the universe was composed of four basic elements of earth, water, air and fire.

The one exception to the materialists' view of a material organizing principle for the universe was Pythagoras. He was an ethicist, a mystic, a spiritualist, the leader of a cult. He believed that the universe was not material but was made up of numbers, which for him were whole numbers. Mathematics was very much in his mind but the numbers of which the universe was composed had to be whole rational numbers. There is a legend that a member of a Pythagorean society was thrown overboard on a voyage for revealing to a member of the crew the darkly held secret that the length of the hypotenuse of a right triangle whose sides were each of unit length could not be expressed as the ratio of two integers. This fact, which was so disturbing to the Greek mathematical mind, posed no problem to the Hindus.

The ancient Greek philosophers might have derived the idea of a unifying principle guiding the universe from the Hebrews' notion of one God, Jahweh, the creator of the universe. The Hebrews developed a notion of causation and the prime cause which they attributed to Jahweh. In fact, the name Jahweh incorporates the notion of causing to come into being. "The enigmatic formula in Ex. 3:14, which in the Biblical Hebrew means 'I am what I am,' if transposed into the form in the third person required by the causative Jahweh, can only become 'Jahweh asher jihweh (later jihyeh), 'He Causes to Be What Comes into Existence.'" The Hebrew notion of causation did not develop along the logical scientific lines of the Greeks. Rather, they incorporated it into their unique sense of history, with its promise of the future, the Promised Land, and their role as the Chosen People (Albright 1957, 87).

A very early notion of evolution was developed by Anaximander who believed that the origin of life was in the sea after which it moved onto land. He suggested that humans were descended from fish. He argued that since the human infant unlike all other animals needs to be nurtured in order to survive that humans must have descended from some other animal. Anaximander's student Xenophanes of Colophon developed his theory of evolution further and made use of fossils he collected to support his conjectures. He also explained meteorological phenomena in terms of clouds and their movement. Another supporter of the notion of evolution was Empedocles who developed a crude notion of natural selection (Everson 2007, 5).

The Paradoxes of Parmenides and Zeno and the Disservice of Logic: Too Much of a Good Thing

Despite the early encouraging signs of a world view based on mathematics and empirical science, the work of Parmenides, who made the first systematic self-consistent use of logic to argue that nothing changes forced Greek thinkers to choose between trusting their senses and trusting logic. As we will see they chose logic, which in our opinion was a disservice to the evolution of scientific thinking. It was not that Greek science was 2000 years ahead of the eventual scientific of the Renaissance it was that the excessive use of deductive logic retarded scientific progress for 2000 years.

The story begins with Heraclitus who believed in constant change and that everything is in flux, summarized by his statement that one cannot step twice into the same river for ever more water flows on. Parmenides, perhaps reacting strongly to Heraclitus' notion of flux set out to show, using logic, that change was impossible. He started with the premise that the notion of non-being was impossible because it is logically self-contradictory. He argued that non-being cannot be. If non-being cannot be, then nothing can change, for if something changes from state A to state B, then A would "not-be". Parmenides argued that since non-being cannot be and once A exists it then cannot not-be and therefore A cannot change because if it were to change then it would not-be which is impossible; therefore, he argued nothing can change.

Parmenides's student Zeno extended Parmenides' argument to show that motion was also not possible. Zeno argued that either a thing is in its place or it is not in its place. But it is impossible for a thing not to be in its place; and therefore nothing moves because if it moved it would no longer be in its place. Another paradox of Zeno is that one can never get to the end of a path because first one had to get half way to the end of the path and then one had to get half way to what remained of the path to be traversed and then one had to get half way to what remained of the path to be traversed and so on and so forth ad infinitum with the result one never gets to the end of the path.

Resolving the Paradox

What is paradoxical of their paradoxical arguments is that they are patent nonsense as common sense tell us that things do change; things move; and people get to the end of their paths. The arguments of Zeno and Parmenides, unfortunately, had a negative effect of the development of Greek science. Their argument, which today would be dismissed as sophistry or logical double-talk, unfortunately for the progress of science, were taken seriously by Greek philosophers. They were not prepared to deal with this new system of rational thought based on logic. Logic was a new technology that purported to provide an infallible way to find truth as Thales demonstrated with his proofs of the Egyptian's empirically derived geometric relationships. Such was the case with the post-Parmenidean scientists and philosophers who, when confronted with the either-or choice between logic and observation, chose logic. Even the empiricist Anaxagoras wrote: "Because of the weakness of our senses we are not able to judge the truth".

Parmenides presented a paradox to Greek philosophy and physics, namely, the contradiction between the commonsense observations of change in the everyday world of experience and his logically consistent arguments against change. Each of the Greek thinkers responded to this challenge by finding a way to resolve this paradox, in each case by carefully inserting into his world view some element that did not change.

Empedocles argued that the four basic substance of earth, water, air and fire never change but only the combinations of them change which explains why we observe change in the world due to the rearrangement of these four basic elements into new combinations. The combinations change but not the four basic elements of which these combinations are composed. Change occurs according to Empedocles by the desire of these four basic elements to form pure states. This explains why heavy things fall back to earth, water flows into water and air and fire rise up. Like materials want to rejoin themselves. The universe according to Empedocles wants to return to an equilibrium position composed of four pure layers of matter consisting of earth, water, air, and fire. This also explains why water is on top of the land, air above the water and the land and fire such as the sun and the stars are in the heavens and flames rise up.

Empedocles' division of the material universe into four basic, immutable, and repeatable elements eventually evolved into the classification scheme of modern chemistry, in which all material substances are described in terms of the approximately 100 immutable elements of which they are composed. An intermediary step in this evolution was that of the alchemists, who added mercury, salt, and sulfur to Empedocles' four elements to describe the material world. The alchemists were on the right track in the way they described nature in terms of elements and compounds. They made the simple error of assuming that gold was an amalgam rather than a pure element.

The desire to resolve the paradox posed by Parmenides and to accommodate change within a static world also led to the formulation of the concept of immutable atoms by Leucippus and Democritus. According to these philosophers, the process of dividing and subdividing matter could not be carried out ad infinitum but would eventually lead to particles they called atoms, which could no longer be subdivided. They were right! The universe is composed of tiny indivisible immutable atoms or particles invisible to the human eye. The Greek word atom literally means "uncut". According to the atomists, there are a finite number of different types of atom, which differ from one another, like the letters of the alphabet, by their size and shape. Each object of the universe is composed of a different combination of atoms just as each word is composed of a different combination of letters. The atomists explained change as the formation of new combinations of atoms. The individual atoms, however, did not change—consistent with Parmenides's idea of immutability. The prediction of atoms some 2500 years before their actual discovery is a tribute to the imagination of the early Greek physicists. The first modern proponents of the atom, Boyle and Dalton, borrowed the idea of the atom directly from them.

The challenge of the paradoxes of Parmenides and Zeno served as a creative spur to the imaginations on Empedocles and the atomists. Their impact on other thinkers was not so positive in that it proved to be destructive to the empirical spirit of Greek science and philosophy. Plato's response was to create two realms: the domain of perceptions where everything is in change and the domain of ideal forms or ideals where all is static and unchanging as Parmenides had suggested. The artificial separation of the empirical and the theoretical domains discouraged the budding empiricism of the Pre-Socratic and Pre-Parmenidean philosophers and reinforced the theoretical and abstract bias of Greek rational thinking.

Aristotle was also influenced by Parmenides and Zeno as he also divided the world into two domains: the imperfect always changing sub-lunar world and the ethereal unchanging heavens. Aristotle's notion of the heavens as an unchanging physical domain distinct from the earth had to be overcome by Copernicus, Kepler, Galileo and Newton some 2000 years later thanks in part to Galileo's use of the telescope and his observation of the craters of the moon and sun spots on the surface of the sun.

Galileo combined his astronomical observations of the heavens with his earth-based experiments in which he carefully measured the behavior of physical objects like balls rolling down an inclined plane and the period of the oscillations of pendulums inspired by his observation of the regular rocking back and forth of a chandelier on the ceiling while sitting in church during Masse. He combined these observations with rigorous mathematical calculations using the data he collected in his experiments.

He wrote, "the universe cannot be read until we have learned the language and become familiar with the characters in which it is written. It is written in mathematical language, and the letters are triangles, circles and other geometrical figures, without which means it is humanly impossible to comprehend a single word. Without these, one is wandering about in a dark labyrinth (Galilei 1623, 171)."

In the end mathematics and science joined up and mutually stimulated each other's development. The phonetic alphabet with which the Greeks transcribed their spoken language provided them with a model for the abstraction, analysis, and classification essential for formulating abstract science. The written word, as transformed by the printing press, also played a key role in the development of modern science as a result of the superior storage and organization of information that print made possible.

The most convincing argument for the importance of writing and numeracy to the origin of science, however, is the simple empirical evidence that all breakthroughs in abstract science have occurred in literate and numerate cultures. As a result of their careful observations, many preliterate societies developed a good understanding of many aspects of nature, such as the medicinal properties of plants or the behavior and habits of animals. What distinguishes their knowledge of nature from abstract science is the way in which they organized their information. The difference between abstract science and primitive science is that the former leads to new discoveries because of the systematic way in which it organizes information. The organizational structures of abstract science naturally pose questions to the scientist which stimulate the process of discovery. Abstract scientists will go out of their way to perform experiments to test the universality of their organizational structures, whereas preliterate cultures are content to describe nature as they encounter it. They also limit their studies of nature to that which is immediately practical to them. The ancient Greeks saw little practical consequence in their first studies of amber, but they pursued their investigations for their own sake, and what they discovered eventually contributed to our present understanding of electricity.

Science as a Language and the Non-Probativity Theorem

To underscore the difference between the mathematical structure of logic and the empirically-based practice of science, which makes use of the language of mathematics we will argue that one cannot prove anything with science. In fact, we will offer a proof of this conjecture. But first let us describe the three forms of logic: deductive, inductive and abductive logic. Deductive logic assumes certain axioms are self-evidently true. The word axiom is derived from the Greek word axiomata, which literally means authority. Then by inference one deduces conclusion, the validity of which is as firm as the axioms one started with. Inductive logic starts with the observations that a certain proposition always seems to be true. One then makes the leap of faith and assumes therefore that the proposition is always true. But this is only a guess. Abductive logic is another form of guess work where one hypothesizes an explanation of why a certain pattern in nature is observed. In a certain sense all science is based on abduction because one can never prove that a scientific proposition is true.

Using deductive logic, we will show that we can prove that if one accepts as an axiom Karl Popper's assertion that a proposition to be scientific it must be able to be falsified that science cannot prove anything to be true. As we have argued, mathematics and science are distinct languages each with its own unique informatic objectives. Mathematics strives to solve equations and to prove the equivalence of sets of statements involving the semantical elements of its language, abstract numbers (such as integers, irrational numbers, imaginary numbers), geometrical objects (such as points, lines, planes, triangles, pyramids, vectors, tensors), sets, operators, etc. A theorem or a proof is a unique syntactical element of the language of mathematics, which we will show cannot be an element of the language of science. A theorem or a proof establishes, using deductive logic, the equivalence of one set of statements, the axioms, whose truths are assumed to be self-evident with the proposition whose truth is to be established by the theorem. The truth of the proposition arrived at using deductive logic depends totally on the truth of the starting axioms.

Science, on the other hand, establishes the veracity of a proposition using the technique of the scientific method of observation, generalization, hypothesis formulation, and empirical verification of the predictions that emerge from the hypothesis. The scientific method is a unique syntactical element of the language of science. In addition to trying to provide an accurate description of nature, science also attempts to describe nature in a systematic manner using the minimum number of elements possible. The description of one phenomenon in terms of another is often claimed to be an explanation. This is one way to interpret this reduction of the number of basic elements needed to describe nature which is a basic goal of science. Science also endeavors to make predictions that can be tested to establish the accuracy of its models. No matter how refined this process becomes and no matter how many reductions and simplifications are made there always remain some irreducible elements that resist explanation or description in terms of simpler phenomena. The process of reduction has to end somewhere. The basic elements in terms of which other phenomena are described can be thought of as the basic atoms or elements of scientific description (MacArthur 2000).

Scientists often makes use of mathematical language to construct their models of nature, especially in the physical sciences. They often employ mathematical proofs to establish the equivalence of mathematical statements within the context of their models. This has led to the popular belief that science can actually prove things about nature. This is a misconception, however. No scientific hypothesis can be proven; it can only be tested and shown to be valid for the conditions under which it was tested. Each proposition must be continually verified for each new domain of observation.

We now turn to a use of mathematical reasoning to show and actually prove that science can never prove the truth of any of its propositions or hypotheses. We need here to clarify what we mean when we use the word truth by distinguishing two types of truth, empirical or verifiable truth and necessary or analytic truth. Empirical truth arises from the matching of a measurement with a model and is always approximate to some degree or other depending on the precision of the measurement and the accuracy of the model. Necessary truth arises out of mathematical reasoning or the use of logic and is exact. Although necessary truth is exact its validity depends totally on the basic axioms from which one starts and which one assumes are self-evidently true. At some point one must rely on belief or faith to establish that an axiom is self-evidently true. The necessary truth of mathematics or logic is therefore artificial. The most one can say about the truth of mathematics and logic is, that subject to the limitations of Gödel's Theorem, it can only demonstrate the equivalence of one set of propositions with another. Mathematics and logic are therefore our first examples of virtual reality. Empirical truth while less precise than necessary truth at least attempts to describe reality. The scientific models are artificial and are only representations of reality but they do have to measure up.

To establish our theorem, the Science Non-Probativity Theorem, we will make use of a basic axiom of the scientific method, namely, that for a statement or an assertion to be considered as a scientific statement it must be tested and testable and, hence, it must be falsifiable. It was Karl Popper (1959) who first formulated the axiom that a proposition has to be falsifiable to be considered as a scientific proposition. If a proposition must be falsifiable or refutable to be considered by science then one can never prove it is true for if one did then the proposition would no longer be falsifiable, having been proven true (in the sense of necessary truth), and, hence, could no longer be considered within the domain of science. We have therefore proven that science cannot prove the truth of anything. Any proof of the truthfulness of a proposition would put that proposition outside the realm of science and place it within the domain of mathematics or logic. And as was pointed out by Stephen Clark, "Not all proofs are ever intended as 'necessitations'. So what counts as 'proof' will vary between disciplines and practices."

Just for fun we have formulated our theorem formally making use of two axioms.

Axiom 4.1 A proposition must be falsifiable to be a scientific proposition or part of a scientific theory.

Axiom 4.2 A proposition cannot be proven necessarily true and be falsifiable at the same time. [Once proven true, a proposition cannot be falsified and, hence, is not falsifiable.]

Theorem A proposition cannot be proven to be true by use of science or the scientific method.

Proof If a proposition were to be proven to be true by the methods of science it would no longer be falsifiable. If it is no longer falsifiable because it has been proven true, it cannot be considered as a scientific proposition and hence could not have been proven true by science. Q.E.D.

In the spirit of the Science Non-Probativity Theorem and our distinction between necessary and empirical truth, we cannot be certain that this line of reasoning is absolutely valid or "true". After all we have just used the theorem, a syntactical element of the language of mathematics to establish a proposition about the language of science. The validity of our conclusion is no greater than that of our starting axioms. Our theorem is not scientifically valid but as a result of mathematical reasoning we have created a useful probe; one that hopefully will lead the reader to some interesting reflections and insights into the nature and limitation of science. The purpose of this exercise was not, as some have suggested, to challenge the usefulness of science or the validity of its methodologies but to clarify the nature of scientific truth and contrast it with the necessary truth of logic.

All that science can do is to follow its tried and true method of observing, experimenting, generalizing, hypothesizing and making predictions then testing its hypotheses and predictions. The most that a scientist can do is to claim that for every experiment or test performed so far, the hypothesis that has been formulated explains all the observations made to date and that all predictions have been validated within experimental errors. Scientific truth is always equivocal and dependent on the outcome of future observations, discoveries and experiments. It is never absolute. We hope these arguments establish that the verification of a scientific proposition through empirical testing or observation is not equivalent to proving the truth of that proposition as some would claim.

Conclusion

In this chapter we have attempted to show the strengths and limitations of science when regarded as a language with its dual role of communication (description) and information processing (predictability). The Non-Probativity Theorem underscores a long held belief that scientific truth is not absolute but always subject to further testing.

Chapter 5
Computing and the Internet: The Six Languages of Speech, Writing, Math, Science, Computing and the Internet

In Chaps. 2–4 we described how spoken language emerged from mimetic communication and then evolved into written language and mathematics and from there into the language of science. Each new language emerged from its predecessors as a way of dealing with the information overload that its predecessors were unable to cope with. Science and science-based technology in turn gave rise to an information overload, which led to the automation of computing mechanically and eventually to computers, mainframes at first followed by minicomputers, personal computers, tablets and smart phones. Computing gave rise to another information overload that gave rise to the Internet and the World Wide Web.

This process of information overload giving rise to a remedy in the form of a new information medium or technique followed by another information overload will continue as long as humans live on this planet and evolve their technology. In this chapter we will study the emergence of computing, the fifth language after speech, writing, mathematics and science and the sixth language of the Internet.

One of the impacts of science over the centuries was the complexification of science-based technology and as a consequence the rapid increase in population. A new level of information overload emerged in the late nineteenth century. The United States government called upon Herman Hollerith who had developed an electromechanical tabulating device to help with the gathering and processing of information for the 1890 US census. Data was entered onto cardboard cards by punching holes in them much like the computer cards of the mainframe computers of the 1950s through the 1970s. The cards were placed on the tabulating machines in which pins were impressed on the cards. Those pins that made it through the holes in the card completed an electric circuit and the information represented by the punched hole was tabulated. The machines were quickly adapted to a number of other commercial activities. In 1911 Hollerith sold "his Tabulating Machine Company to financier Charles Flint for US $2,312,100, and the company became part of Flint's Computing-Tabulating-Recording Company. In the 1920s, C-T-R evolved into IBM (www-03.ibm.com/ibm/history/ibm100/us/en/icons/tabulator)." Once again information overload led to a new development, in this case, automated

R. K. Logan, I. Pruska-Oldenhof, *A Topology of Mind*, Mathematics in Mind,
https://doi.org/10.1007/978-3-030-96436-8_5

tabulating machines and eventually in the 1940s to computers. And as they say the rest is history, a well-documented history.

Mainframe computers evolved into minicomputers and personal computers and from there to networked computers and then the Internet, the World Wide Web, search engines and then smartphones which provided computing with mobility. Along the way each development led to an information overload that stimulated the next development. With personal computers there was so much information to share that the Internet took off. The information overload of the Internet led to the World Wide Web and from there to Web 2.0, social media, search engines and smartphones, i.e. mobile computers equipped with the capability of telephoning, a virtual cascade of online information processing developments.

Computing with its programming languages developed its own unique semantics and its own unique vocabulary in terms of word processing, spread sheets and data bases as examples. Very quickly computing which emerged to deal with the information overload of science, created its own information overload due to the desire of its users to share the information they created with others. The Internet emerged as way of dealing with that overload as another language with its own unique syntax and semantics.

Without computing, natural scientists, engineers and social scientists would not have been able to manage the information overload created by their disciplines. It was only after its initial application as an informatics device that computing was also used for communications, and hence, its name in English is "computer" (as in calculator) and not "word processor," even though far more users process words with computers than compute or calculate numbers with them.

The Internet is another example of a new form of language which emerged from an information overload. Computing increased the sheer number of messages that needed to be communicated as well as the number of people that needed to be communicated with. As the world shrunk to the dimensions of a global village the number of people in the village with whom one wanted to communicate increased dramatically. This information overload or traffic jam of messaging gave rise to networking, client-server systems and finally the Internet. The Internet as opposed to client server systems was able to embrace the entire global community in a single electronic embrace. As is so often the case a quantitative change created a qualitative change and as a result a new language emerged, the Internet or the sixth language.

Starting with the ability to record ideas through writing and mathematical notation, human thought has become increasingly more complex. The need to model more complex phenomena has driven the development of the six modes of language. Consequently, each new mode of language is informatically more powerful than its predecessors, but at the same time a little less poetic with the exception of the Net which because of its visual and audio elements is able to incorporate the arts and mode of artistic expression. Our model of the evolution of language is one in which the information-processing capacity of language becomes more and more important as the complexity of human thought increases. It is essential to remember, however, that all forms of language possess a dual capacity for communication and information processing. Computers and the Internet are also communication devices and the spoken word has an informatics capacity.

The language of computing includes all of the semantical and syntactical elements of the earlier four modes of language. It also possesses its own semantical and syntactical elements by virtue of the activities of both its programmers and its end users. The semantics of the programming languages and end-user software programs specify computer inputs and outputs. The syntactical structures of programming languages and end users' software formalize the procedures for transforming inputs into outputs. These syntactical structures are basically unambiguous algorithms for ensuring the accuracy and the reliability of the computer's output. The syntactical structures that arise in a programming language or a relational database differ from the other language modes so that the user can take advantage of the computer's rapid information-processing speeds.

Although the Internet and the World Wide Web incorporate all of the semantical and syntactical elements of computing they also include their own unique elements for both categories. Perhaps we should clarify the relationship between the Web and the Net. The World Wide Web is one of many different elements of the Internet which include its email facilities, listservs, chat rooms, ftp facilities, Telnet facilities, Web pages, Web sites, intranets, extranets, portal sites and e-commerce sites. Each of these facilities represent the semantical elements of the sixth language of the Internet. Listservs are a way of distributing emails to a group of users who share an interest in a common topic which allows an asynchronous email dialogue to take place. A chat room like Google hangout is a place on the Internet where people can meet online to discuss a topic in real time. FTP (file transfer protocol) allows files to be transferred from one computer to another through the medium of the Internet. The Telnet facility allows a user to access their server or home computer from anywhere in the world as long as they can find access to the Internet. Web pages are components of Web sites or intranets which integrate text, graphics, video and audio. Web sites are publicly accessible collections of Web pages that can be found using a URL or Internet address. Intranets are private Web pages that can only be accessed by qualified users. Extranets are a collection of Web sites that can be accessed from a single Internet address. An e-commerce site is a Web site that allows commercial transactions to take place using credit cards or digital money.

The Internet has a number of unique syntactical elements. One of the unique syntactical elements of the sixth language is hypertext which makes it possible to link all of Web sites and Web pages in cyberspace to form one huge global document. Another unique syntactical element is the Internet Protocol which allows all of the computers connected to the Internet to form one huge Global Network and makes the Web, ftp and telnet all possible. McLuhan's prediction of a Global Village has been realized. Still another unique syntactical element of the Internet are the search engines which increase access to knowledge and information and hence provide an extra level of communication that the other forms of verbal language cannot match. The search engine also facilitates people finding each other and hence contributes to the creation of a global knowledge community.

The Advent of Computing

The computer may be regarded simply as a device for the automated processing, storing, retrieving, organizing, and communicating of information. It basically manipulates abstract symbols, whether they are words, numbers, equations, or databases. Computers change inputs into outputs in a systematic, reliable, and relatively rapid manner. The operations of the computer are closer to its French name, ordinateur, or "that which orders," than its English name, which better describes its humble origins as an automatic calculator. To understand its implications, however, it is useful to define computing as a language in the way speech, writing, mathematics, and science are defined as language.

By defining computing as a language we are making a distinction between the computer, which is a human build hardware artifact, and the processes and skills needed to use this hardware to organize and communicate information. It is the organizing and communicating of information and ideas which is the language of computing and the computer which is the medium. The same parallel holds for speech, writing and mathematics. The medium of speech consists of the vocal cords, the tongue of the speaker, the air to transmit sound waves and the ear to receive them. For writing and math the medium is the pen, the ink, the hands that write and the eyes that read. The language in these three examples are the spoken words, the written words and the mathematical symbols which organize and communicate information, ideas and thoughts.

The language of computing may be considered to include all of the information resident in computers, as well as all of the information and techniques needed to operate computers. The information resident in computers may be regarded as the content of the language of computing, and includes spoken language, written language (literature), mathematics, and science. The techniques for using computers may be seen as the grammar of the language of computing. The reason that computing may be considered as another level of language is that its strategies for processing and communicating information are very different from that of the other modes.

Just as literacy and numeracy were essential for the development of abstract science, so in turn was modern science essential for the development of computers. Modern science provided the necessary technical ingredients such as electronics, circuitry, the magnetic storage of information, mathematics, logic, Boolean algebra, solid state physics, and telemetry. Science also provided the motivation to design computers because progress in a number of fields required computational techniques for calculating complex mathematical problems which could not be solved using standard numerical techniques. The computer, with its ability to perform large numbers of simple calculations at high speed, provided the solution to these problems. Complex calculations could be broken down into a large number of smaller and simpler problems using the techniques of numerical analysis. The solutions to the simpler problems could be calculated by developing algorithmic procedures which in turn could be automated. This is the basic principle behind the computer-based solutions of the complex mathematical problems that arise in the domain of science, as well as governmental administration (for example, collecting taxes and tabulating census results) and big business (for example, maintaining airline schedules or financial records).

The Advent of the Internet

Originally, the Internet was organized by the American military and designed to provide a backup communications system if the United States lost large blocks of its normal communications system due to a nuclear attack by the Russians. As a result, the Internet was designed without a central switching system or headquarters. It is a distributed network where each of the computers on the Internet contributes to its communication and distribution capability. The Internet makes use of the infrastructure of the telephone and more recently the cable companies but it is not operated by any telephone company or other form of central administration. It is the world's largest self-policing community of individuals and organizations operating without any centralized regulating authority. The lack of a regulating authority or an agency to vet its content have implications for the role the Internet plays in society, particularly in the education of the young. This is an issue we will address later on in our discussion.

The Internet may be regarded simply as a global network of computer users and servers that are linked using the TC/ICP protocol and the infrastructure of telecommunication lines that span the planet. The Internet permits the sharing of information in both synchronous and asynchronous modes in a variety of formats including text, graphics, video and audio. To understand its implications, however, it is useful to define the Internet as a language in the way speech, literature, mathematics, science, and computing are defined as languages. Once again we make a distinction between the hardware artifacts of the computers, servers, routers, and telecommunications links which compose the medium of the Internet and the process of organizing and communicating information which is the language of the Internet.

The language of the Internet may, therefore, be considered to include all of the information resident in computers across the planet that can be accessed by the Internet as well as all of the information and techniques needed to operate the Internet. One may consider local dialects of the Internet to be all the information contained on local intranets that are only accessible to a closed group of users as well as all of the information and techniques needed to operate the intranet. The information resident in the Net (or intranets) may be regarded as the content of the language of the Net (or intranets), and includes text, graphic, video and audio. The techniques for using the Net (or intranets) may be seen as the grammar or syntax of the language of the Net (or intranets) which include the TC/ICP protocol, browsers such as Netscape and Explorer, and markup languages such as HTML for hypertext and XML for e-commerce applications. The reason that the Net (or intranets) may be considered as another level of language is that its strategies for processing and communicating information are very different from that of the other five modes of language. Although the language of computing is absolutely necessary for the development of the Internet the way in which it has developed and been use is completely different from non-networked computers. The Internet is a hybrid medium of computing and telecommunications.

The Mind in Mathematics

The focus of this study is the Mathematics in Mind, the name of the series for which this book was written. We, however, want to make a slight detour and talk about the flip side of this theme, namely the mind in mathematics, as is claimed by the proponents of strong AI (artificial intelligence), artificial general intelligence (AGI), the Singularity or the technological singularity. The idea of the Singularity is that someday a programmer will be able will design a computer with an AI or AGI capability that will allow it to design a computer even more intelligent than itself and that computer will be able to do the same and by a process of iteration a technological singularity point will be arrived at where post-Singularity computers will be far more intelligent than us poor humans who only have an intelligence designed by nature through natural selection and evolution. At this point according to some that embrace this idea, the super-intelligent computers will take over and we human will become their docile servants.

Basically, the proponents of the technological Singularity believe that a mind can be created using the mathematical skills of programming and hence they believe a mind in mathematics could come into being.

As this is a detour from our original mission to study mathematics in mind we will limited ourselves to making two important points. First, we do not believe that the technological Singularity can be achieved simply because a computer is nothing more than a series of logic gates and the human mind makes use of logic gates but it is so much more. We, in fact, believe the idea of the technological Singularity is a dangerous idea because it could lower our standards as to what constitutes human intelligence and actually make it possible for us to become the servants of our own technology.

What makes the idea of the technological Singularity so ridiculous is that human intelligence is not just a matter of logic and rationality but that it also entails the following characteristics of what makes us human, namely having purpose, objectives, goals, intuition, imagination, humor, emotions, passion, desires, curiosity, values, morality, experience, wisdom, and judgement. As well as being capable of experiencing pleasure, beauty, and joy. None of these attributes can be possessed by a computer. Computers can only deal with symbolic information but human intelligence is not limited to dealing with symbolic information. Human intelligence and expertise depends on unconscious instincts and not just conscious symbolic information.

The human mind and its thought processes are emergent, non-reductionist phenomena. Computers, on the other hand, operate making use of a reductionist program of symbolic manipulations. AGI is linear and sequential whereas human thinking processes are simultaneous and non-sequential. Imagination, curiosity and creativity lie outside of and defy logic, whereas a computer is bound by logic.

Humans experience a wide variety of emotions some of which motivate art and science. Emotions, which are a psychophysical phenomenon, are closely associated with pleasure (or displeasure); passion; desires; motivation; aesthetics and joy. Every human experience is actually emotional. It is a response of the body and the brain. Every experience is about what action to take or as Terry Deacon suggested in a private communication, "acting to do it again or not do it again."

Emotions play an essential part in human thinking as neuroscientist Antonio Damasio has shown:

> Damasio's studies showed that emotions take [or play] an important part in the human rational thinking mechanism (Martınez-Miranda and Aldea 2005, 326).
>
> For decades, biologists spurned emotion and feeling as uninteresting. But Antonio Damasio demonstrated that they were central to the life-regulating processes of almost all living creatures. Damasio's essential insight is that feelings are "mental experiences of body states," which arise as the brain interprets emotions, themselves physical states arising from the body's responses to external stimuli. (The order of such events is: I am threatened, experience fear, and feel horror.) He has suggested that consciousness, whether the primitive "core consciousness" of animals or the "extended" self-conception of humans, requiring autobiographical memory, emerges from emotions and feelings (Potin 2014).

Terrence Deacon (2012, 512 and 533) in *Incomplete Nature* also claims that emotions are essential for mental activities:

> Emotion… is not merely confined to such highly excited states as fear, rage, sexual arousal, love, craving, and so forth. It is present in every experience, even if often highly attenuated, because it is the expression of the necessary dynamic infrastructure of all mental activity… Emotion… is not some special feature of brain function that is opposed to cognition.

Computers are incapable of emotions which in humans are inextricably linked to pleasure and pain because they have no pain nor any pleasure and hence there is nothing to get emotional about. In addition, they have none of the chemical neurotransmitters and the drives associated with them, which is another reason why computers are incapable of emotions. Without emotions computers lack the drive that are an essential part of intelligence and the striving to achieve a purpose, an objective or a goal. Emotions play a key role in curiosity, creativity and aesthetics, three other factors that are essential for human intelligence.

Singularitarians are essentially dualists that embrace the dualisms between body and mind and between reason and emotion. They are the last of the behaviorists who have replaced the Skinner box with a silicon box (today's computers). The mind is not just the brain and the brain is not just a network of neurons operating as logic gates. The human mind extends into the body, is extended into our language (Logan 2007) and extended into our tools (Clark and Chalmers 1998; Clark 2003).

Curiosity

I have no special talent. I am only passionately curious.—Albert Einstein

Curiosity is both an emotion that motivates behavior. But given that computers are not capable of emotions they cannot be curious and hence lack an essential ingredient for intelligence. Curiosity entails the anticipation of reward, which in the brain comes in the form of neurotransmitters like dopamine and serotonin. No such mechanism exists in computers and hence they totally lack native curiosity. Curiosity if it exists at all would have to be programmed into them. In fact, that is exactly what NASA did when it sent its Mars rover, aptly named Curiosity, to explore the surface of Mars.

Curiosity and intelligence are highly correlated. Advances in knowledge have always been the result of someone's curiosity. Curiosity is a characteristic that only a living organism can possess and no living organism is more curious than humans. How could a computer create new forms of knowledge without being curious? But that level of curiosity would have to be the curiosity of the programmers who create the AGI creature. And since the curiosity programmed into the AGI device cannot exceed native human curiosity this represents a real barrier to the achievement of intelligence by the technological Singularity.

Values and Morality

Because a computer has no purpose, objectives or goals it cannot have any values as values are related to one's purpose, objectives and goals. As is the case with curiosity, values will have to be programmed into a computer and hence the morality of the AGI device will be determined by the values programmed into it and hence the morality of the AGI device will be that of its programmers. This gives rise to a conundrum. Whose values will be inputted and who will make this decision, a critical issue in a democratic society.

Decision Making, Experience, Judgement and Wisdom

The only source of knowledge is experience.—Albert Einstein

If the challenges of programming an AGI device with a set of values and a moral compass that represents the will of the democratic majority of society, there is still the challenge of whether the AGI device still has the judgment and wisdom to make the correct decision. In other words, is it possible to program wisdom into a logical device that has no emotions and has no experiences upon which to base a decision. Wisdom is not a question of having the analytic skills to deal with a new situation but rather having a body of experiences to draw upon to guide one's decisions. How does one program experience into a logic machine?

Intelligence requires the ability to calculate or compute but the ability to calculate or compute does not necessarily provide the capability to make judgments and decisions unless values are available, which for an AGI device requires input from a human programmer.

Another dimension of knowledge related to experience that must be taken into account is the relationship between tacit knowledge and explicit knowledge so important for collaboration.

> One way to describe tacit knowledge is in terms of intuition. Tacit knowledge is personal, intuitive knowledge, whereas explicit knowledge is the kind of knowledge that can be learned from a book. There is a vast difference between book learning—explicit knowledge and experienced-based learning—tacit knowledge.
>
> Intuition is defined in *Webster's New World Dictionary* as "the immediate knowing of something without the conscious use of reasoning." There are times when we know something to be true but we do not know why or how we arrived at the understanding. One form of intuition arises because we know something so well and so thoroughly that we do not have to reason things out again but we immediately know it. This is tacit knowledge, as opposed to, explicit knowledge which is "formal and systematic" (Nonaka and Takeuchi 1995). Tacit knowledge according to Nonaka and Takeuchi is "deeply rooted in an individual's action and experience" (ibid., p. 8). It is intuitive and subjective whereas explicit knowledge is scientific and objective (Logan and Stokes 2004, 40).

A computer can only deal with explicit knowledge as it has no experiences. Knowledge and learning only advance through collaborations. AI has many useful applications but collaboration and the exchange of tacit knowledge is not one of them.

A Partnership of Mathematics in Mind and Mind in Mathematics

The reader should not construe the critique of the technological Singularity as a critique of AI. An increase in human intelligence will not come from strong AI or the technological Singularity but AI can be a great ally of human intelligence. Just as the six media of speech, writing, mathematics, science, computing and the Internet have each had a positive impact on the increase of human intelligence AI will have the same kind of positive impact. AI will help solve some problems that can be formulated as purely logical exercises freeing the human mind for more creative work. Human intelligence has thrived in situations where groups of thinkers could form a collective form of intelligence. Universities where scholars could come together to collaborate are one example of a collective form of intelligence. A business corporation is another environment where intellectual collaborations has thrived. With networked computer technology a collective mind with a global scale is emerging, which one might consider as a form of artificial intelligence or perhaps more appropriately as collective intelligence.

Conclusion

Computing and the Internet, products of the mathematics in the minds of humans, are tools that have expanded the human mind into new territories and new adventures that the minds of 500 years ago or even 100 years ago could never have imagined.

Prologue to Part II Spiral Patterns in Nature and Human Thought

In Part II we will examine the spiral patterns in the abiotic physical universe, the biosphere, in human thought and in human artistic expression. We believe that spirals are everywhere and are the fundamental part of nature and human cognition, as has been suggested by all the scholars, authors and artists we cover in this section of our book. We begin in Chap. 6 with a mathematical description of spirals and helixes. In Chap. 7 we describe the spiral structures of the physical abiotic world of galaxies, planetary movements, electromagnetic waves, whirlpools, hurricanes and tornados providing pictorial representations of these physical phenomena. In Chap. 8 we turn to biology and describe the spiral structures found in living organisms providing pictorial representations of these biological specimens. In Chap. 9 we focus on the spiral structures found in history and philosophy. In Chap. 10 we focus on the spiral structure of the thinking of Marshall McLuhan, a mentor whose spiral thought patterns motivated our study of spiral patterns. In Chap. 11 we examine the spiral structures in the fine arts.

In Chaps. 9–11 we will follow the spiral course of the thoughts set out by the cultures, authors and artists that we reference. We will demonstrate how this form manifests itself differently in each of the examples that we consider. Most importantly, its focus is on spiral as a dynamic form and its creative expression; therefore, selection of works for Chaps. 9–11 was determined by the attention that those works accorded to the dynamic form of the spiral. What interests us is the hidden ground, the structure or pattern of art and thought both scientific and poetic, which we believe is the spiral.

McLuhan's understanding of media and his ecological approach will inform our approach in studying the universal spiral structure. We regard the spiral as a medium and through this study we hope to discover or decipher its message. Our initial hypothesis is that spiral structures involve both progression and regression as well as feedforward and feedback. In most cases the overall trend of the spiral structure is one of increase and growth and sometimes an increase in complexity. There are situations, however when the spiral structure can represent a decline. The dynamics of the spiral is that of a cycle in that the spiral returns to its starting point but in the meantime, there has been growth or decline in a different dimension.

Hector Sabelli and Louis Kauffman (2013) in an article entitled "The Biotic Logic of Quantum Processes and Quantum Computation" described why the spiral structure play such an integral part of so many phenomena in nature. They wrote, "In nature many changes are cyclical—as in day/night, breathing in/breathing out, first one side dominates, then the other—but these cycles do not come back exactly to where they started; they don't make a perfect circle. Instead, change occurs in spiral fashion."

Reference

Sabelli, Hector and Louis H. Kauffman. 2013. "The Biotic Logic of Quantum Processes and Quantum Computation." In Franco Orsucci and Nicolette Sala (eds) Complexity Science, Living Systems, and Reflexing Interfaces: New Models and Perspectives. Hershey PA: IGI Global. Chapter 7, 112–183.

Chapter 6
The Mathematical Structure of Cyclic Phenomena: Spirals, Helixes, Revolutions, Waves and Oscillations

In Australian Aboriginal and Maori literature, the circle and the spiral are the symbolic metaphors for a never-ending journey of discovery and rediscovery.

Eva Knudson (2004)

Introduction

The cyclic structures and motions of spirals, helixes, revolutions, waves and oscillations are universal forms and enduring archetype representing the nature of inorganic matter ranging in scales from galaxies and the universe as a whole to atoms and electromagnetic radiation, various aspects of life in the biosphere, the creative and symbolic processes of human culture and, for some such as Vico and Nietzsche, of history.

The spiral structure is a ubiquitous form and universal that one finds in nature in physical structures like whirlpools, tornados and hurricanes here on Earth and galaxies throughout the universe (see Chap. 7) and in biological life forms from the simplest viruses and bacteria all the way up to our own human species such as the DNA (see Fig. 6.16) in their genetic makeup that determines their development, growth and behavior and in the patterns of biological forms like the florets of a sunflower, the petals of many flower, the shell of the Nautilus mollusk and the branches of trees (see Chap. 8). The spiral structure is also found in the following domains of human study and cultural expression: media, the evolution of technology, literature, history, philosophy, architecture and art.

The spiral structure in a two-dimensional plane is constantly increasing in the angular direction whereas its distance from the center of the plane is either increasing or decreasing as shown in Figs. 6.2, 6.5 and 6.7. The three-dimensional spiral is constantly reversing direction in one plane and advancing or retreating in the dimension orthogonal to that plane as is the case of the helix structure shown in Fig. 6.16.

The spiral involves both progression and regression as well as feedforward and feedback. In most cases the overall trend of the spiral structure is one of increase

R. K. Logan, I. Pruska-Oldenhof, *A Topology of Mind*, Mathematics in Mind,
https://doi.org/10.1007/978-3-030-96436-8_6

and growth and sometimes an increase in complexity. There are situations, however when the spiral structure can represent a decline. The dynamics of the spiral is that of a cycle in that the spiral returns to its starting point but in the meantime, there has been growth or decline in an orthogonal direction. The word spiral derives from Middle French circa 1550 which in turn derives from Medieval Latin spiralis "winding around a fixed center, coiling" (mid-13c.), and then from Latin spira "a coil, fold, twist, spiral," and finally from Greek speira "a winding, a coil, twist, wreath, anything wound or coiled," from the base sper- 'to turn, twist' (http://www.etymonline.com/index.php?term=spiral).

Other forms of cyclic motion or structures are waves, oscillations, vibrations and orbital revolutions. Waves occur in many different media. There are the waves in the oceans and lakes of the world. There are sound waves, which are mainly compressions waves in the atmosphere but are sometimes transmitted in water. There are electro-magnetic waves which are basically oscillations of the electric and magnetic fields.

The spiral as a geometric form (see Figs. 6.2, 6.3, 6.4, 6.5, 6.6, 6.7, 6.8, 6.9, 6.10, 6.11, 6.12 and 6.13 in this chapter) can be represented as an arithmetic series as was done by Fibonacci. It is also a form that can be found at a scale as large as our Milky Way galaxy (see Figs. 8.2 and 8.3 in Chap. 8); and a scale as small as electromagnetic waves such as gamma rays, X-rays, light, and all other forms of electromagnetic radiation. And spirals can be found at all the scales in between such as cyclones and hurricanes (see Figs. 7.11 and 7.12 in Chap. 7), tornados (see Fig. 7.13 in Chap. 7), whirlpools (see Figs. 7.8 and 7.9 in Chap. 7), nautilus mollusk shells (see Fig. 6.10 below) and sunflower seeds (see Fig. 6.6 below).

Spirals can occur in both space and time. Spatial spirals, which occur in the res extensa of the material world, are easily visualized as in Figs. 6.6, 6.10, 6.14 and 6.16 below. But spirals also occur in time in the res cogitans of the symbolic world of human culture. TS Eliot suggests in Burnt Norton the idea of spirals in the time dimension through the simultaneity of past, present and future as described in the passage from Burnt Norton that we opened this chapter with. Other examples include the cyclic patterns of history and culture, the philosophy of Vico and his idea of ricorso; Hegel's dialectic structure, Marx's dialectical materialism, Nietzsche's and Eliade's notion of eternal return or eternal recurrence, the psychology of Freud, the communication studies of McLuhan and his Laws of Media and his notion of the reversal of cause and effect (see Fig. 6.1 for images of Vico, Hegel, Nietzsche and McLuhan).

We will also encounter in Chap. 11 the spiral structures in the following areas of artistic expression:

- arts and letters such as Petrarchan sonnets, Dante's descent into the Inferno, the literature of Proust and James Joyce's Finnegan's wake,
- the art movement of vorticism,
- the music and fugues of Bach and other composers,
- cinema such as Sorel Etrog's film Spiral,
- the architectural form of the golden spiral and the spiral staircase and much much more.

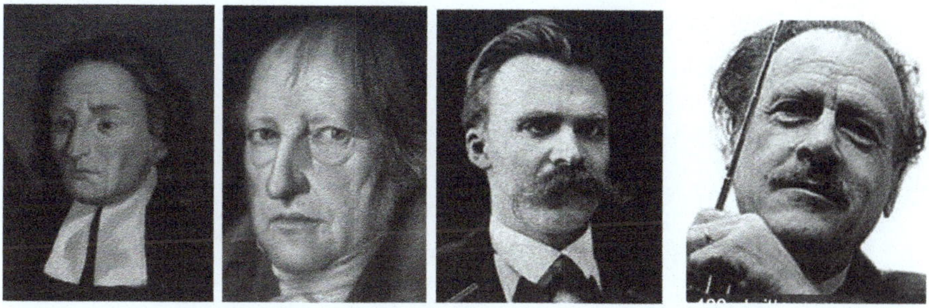

Fig. 6.1 Images of Vico, Hegel, Nietzsche and McLuhan from left to right respectively

The spiral structure connects fields of study, past, present and future, in the arts and the sciences. The spiral structure appears in the feedforward and feedback of cybernetics and human culture where art is feedforward and science and engineering are feedback. The spiral structure entails transformation, a process and a pattern that is always changing, evolving and involving and growing. Of course, the spiral also connotes in some situations destruction as in the expression "spiraling out of control." But even here one can put a growth spin on this expression by suggesting that spiraling out of control indicates the growth of destructive forces. The Hindu God Shiva, the creator and the destroyer, represents the two aspects of the spiral, the spiral of growth and the spiral of destruction. We shall also encounter upward or outward spirals and downward or inward spirals. Some spirals are two-dimensional as those shown in Figs. 6.7, 6.8 and 6.9 and some are three-dimensional like DNA shown in Fig. 6.16. Three dimensional spirals are sometimes referred to as helices or coils.

Mathematical Spiral Structures

There are two simple ways to describe a spiral structure mathematically, one arithmetically and the other geometrically. The spirals in this section are represented by an algebraic formula and by geometric figures. The images in this section are courtesy of the Wikimedia Commons.

The Spiral of Archimedes

The way to generate the Spiral of Archimedes structure is to plot the formula $R = a\theta$ on polar graph paper (Fig. 6.2):

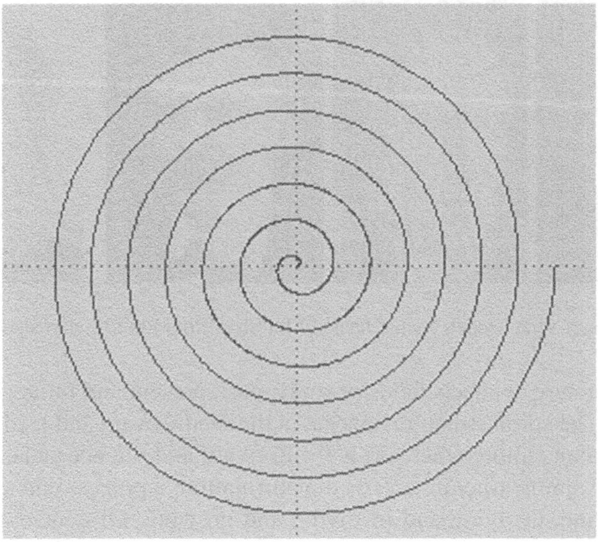

Fig. 6.2 Courtesy Wikimedia Commons

The Euler Spiral

The Euler spiral (see Figs. 6.3 and 6.4) also known as the Cornu spiral or clothoid is a curve whose curvature changes linearly with its curve length, where the curvature of a circle is defined as 1/R where R is the radius of the circle. For a circle with a small radius the curvature is great and for a large radius the curvature is gentle. In the limit as the radius approaches infinity the curvature approaches zero, which is the curvature of a straight line. The Euler spiral is used in railroad and highway design to reduce the amount of the centripetal force acting on the train or vehicle on curves in the track or roadway.

Fermat's Spiral

Fermat's spiral is generated in polar co-ordinates by the radius distance r and the angle θ by the formula r = square root (θ) (Figs. 6.5 and 6.6).

The Hyperbolic Spiral

The hyperbolic spiral is generated by the formula r = a/θ (Fig. 6.7).

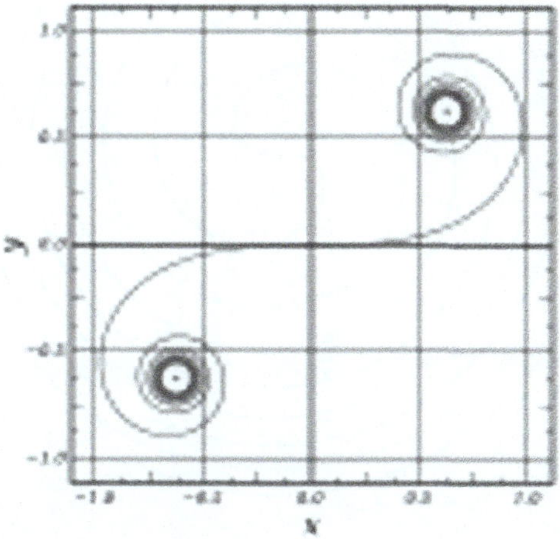

Fig. 6.3 Courtesy Wikimedia Commons

Fig. 6.4 Courtesy Wikimedia Commons

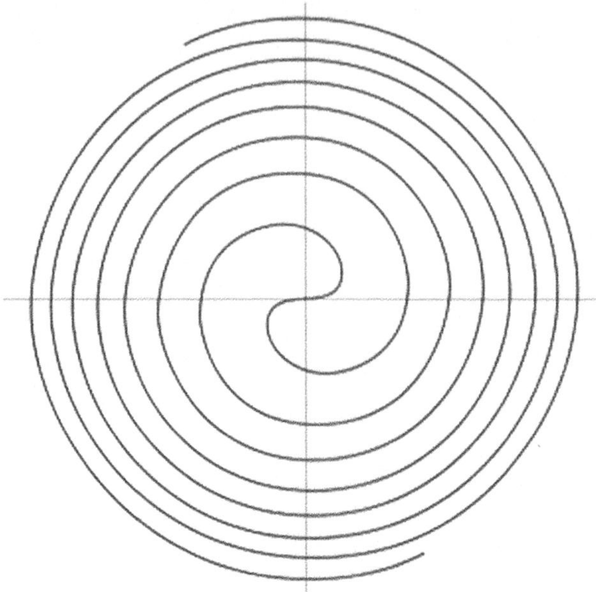

Figs. 6.5 and 6.6 Fermat Spiral plotted on graph paper and found in the arrangements of the florets on the head of a sunflower follow. (Both images courtesy Wikimedia Commons)

Fig. 6.5 (continued)

The Lituus

The Lituus is a spiral that is generated by the formula r = 1/square root (θ) (Fig. 6.8).

The Logarithmic Spiral

The logarithmic spiral is generated by the formula $r = a\ e^{b\theta}$ (Fig. 6.9).

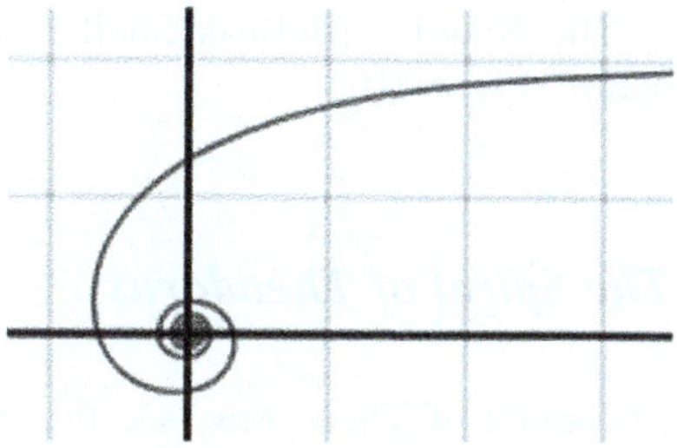

Fig. 6.7 Courtesy Wikimedia Commons

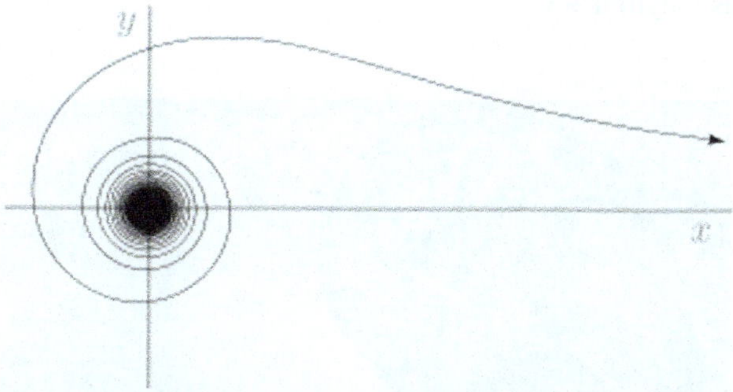

Fig. 6.8 Courtesy Wikimedia Commons

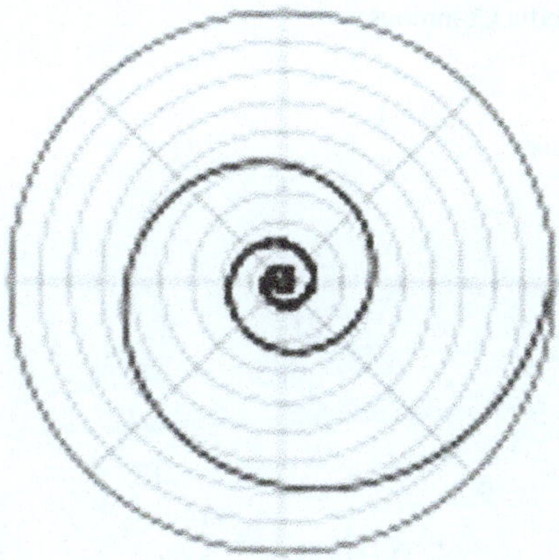

Fig. 6.9 Courtesy Wikimedia Commons

The Nautilus Mollusk Shell is an example of the logarithmic spiral found in nature (Fig. 6.10).

The Spiral of Theodorus

The spiral of Theodorus (see Fig. 6.11) is an approximation of the Archimedean spiral composed of contiguous right triangles where one leg has length 1 and the nth hypotenuse has length n + 1.

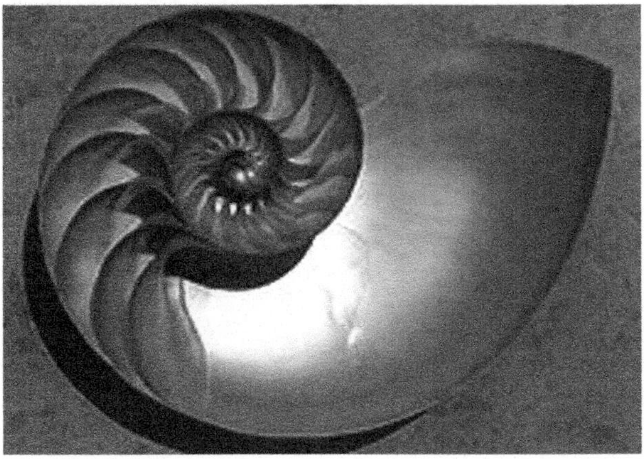

Fig. 6.10 Courtesy Wikimedia Commons

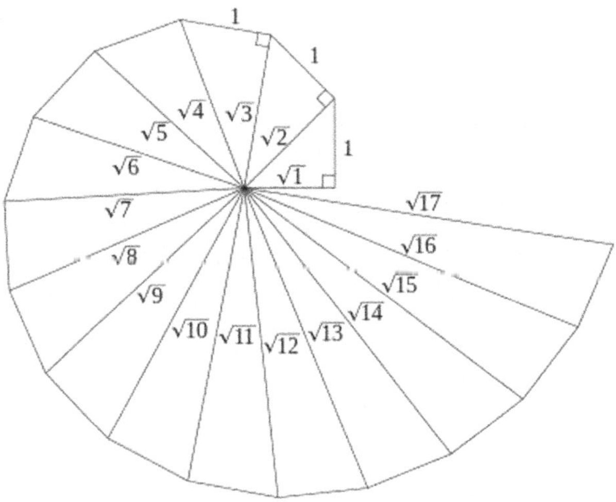

Fig. 6.11 Courtesy Wikimedia Commons

Note that these two-dimensional spirals with the exception of the Euler spiral define a curve on a plane that winds around a fixed center point at a continuously increasing or decreasing distance from that point. Spiral can spiral out or spiral in depending on whether the radius increases or decreases as the spiral turns about a fixed point.

There is also a class of three-dimensional spirals known as helices as is the case of the structure of DNA as shown in Fig. 6.16, which is a double helix. An example of a single helix is the threads on a screw, the handrail on a spiral staircase or a spiral ramp like the one at the Guggenheim Museum is New York City.

Fibonacci Sequence

The Fibonacci sequence of numbers is 0, 1, 1, 2, 3, 5, 8, 13, 21, 34, 55, 89, 144, 233… can be used to generate a spiral known as the Golden Spiral. The Fibonacci sequence is obtained by starting with 0 and 1. Each number in the sequence is the sum of the two numbers that precede it. So 0 + 1 = 1 and 1 + 1 = 2 and 1 + 2 = 3 and 2 + 3 = 5 and so on and so forth. We can represent the Fibonacci sequence with an algebraic formula. Let x_n be the value of the nth term in the sequence then $x_{n+2} = x_{n+1} + x_n$, where $x_1 = 0$ and $x_2 = 1$. Many structures in nature as we will discover follow the Fibonacci sequence. A hint of why this is the case can be gleaned by looking at the way Fibonacci generated his sequence of numbers in 1202 by considering how mating pairs of rabbit would propagate if a new born single pair of breeding rabbits was placed in a field and one were to assume that this pair would be able to mate after 1 month and produce breeding pairs once a month. One also has to assume that all subsequent issues of the original pair would also be able to mate after 1 month and would produce breeding pairs once a month. At the beginning there would be one breeding pair and at the end of the first month there would still be one pair but after 2 months there would be two pairs, and after the third month 3 pairs and then after each subsequent month the number of pairs would follow the Fibonacci sequence. This explains why many of the structures in the biological world follow the Fibonacci sequence.

The Golden Spiral

We can generate the Golden Spiral by placing squares side by side where the side of each square corresponds to the next number in the Fibonacci sequence, which is a tiling with squares whose side lengths are successive Fibonacci numbers (see Figs. 6.12 and 6.13; the Image on left is from https://www.mathsisfun.com/numbers/fibonacci-sequence.html).

Fig. 6.12 Courtesy www.mathisfun.com

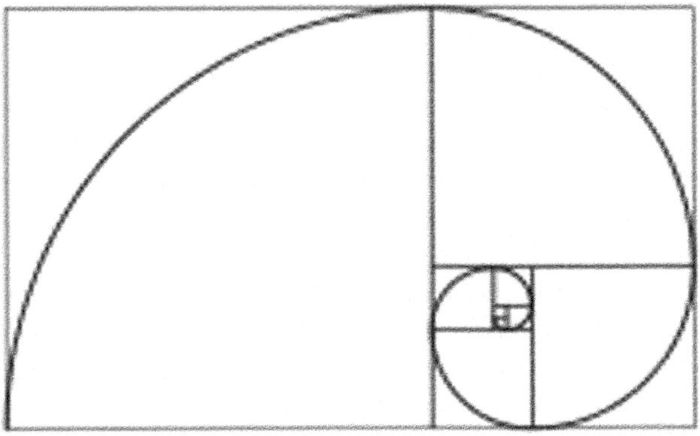

Fig. 6.13 Courtesy Wikimedia Commons

Fractals and Spirals

A fractal is defined as a self-similar structure and some spirals are fractal like in that they have self-similar structures, the logarithmic spiral being one. The vegetable Romanesco broccoli seen in Fig. 6.14 has both a fractal and a logarithmic spiral structure:

Image (Fig. 6.15) of a section of the Mandelbrot set that contain fractal spirals.

Examples of images of physical spirals

http://www.miqel.com/fractals_math_patterns/visual-math-natural-fractals.html

Helixical Structures: A Three-Dimensional Spiral

A helix is the three-dimensional analogue of the two-dimensional spiral

Fig. 6.14 Romanesco broccoli. (Courtesy Wikimedia Commons)

Fig. 6.15 Courtesy Wikimedia Commons

> A **helix** (pl: **helixes** or **helices**) is a type of smooth space curve, i.e. a curve in three-dimensional space. It has the property that the tangent line at any point makes a constant angle with a fixed line called the *axis*. Examples of helices are coil springs and the handrails of spiral staircases. A "filled-in" helix—for example, a "spiral" (helical) ramp—is called a helicoid. Helices are important in biology, as the DNA molecule is formed as two intertwined helice, and many proteins have helical substructures, known as alpha helices. The word *helix* comes from the Greek word ἕλιξ, "twisted, curved" (https://en.wikipedia.org/wiki/Helix).

Here is an image of DNA, a double helix (Fig. 6.16):

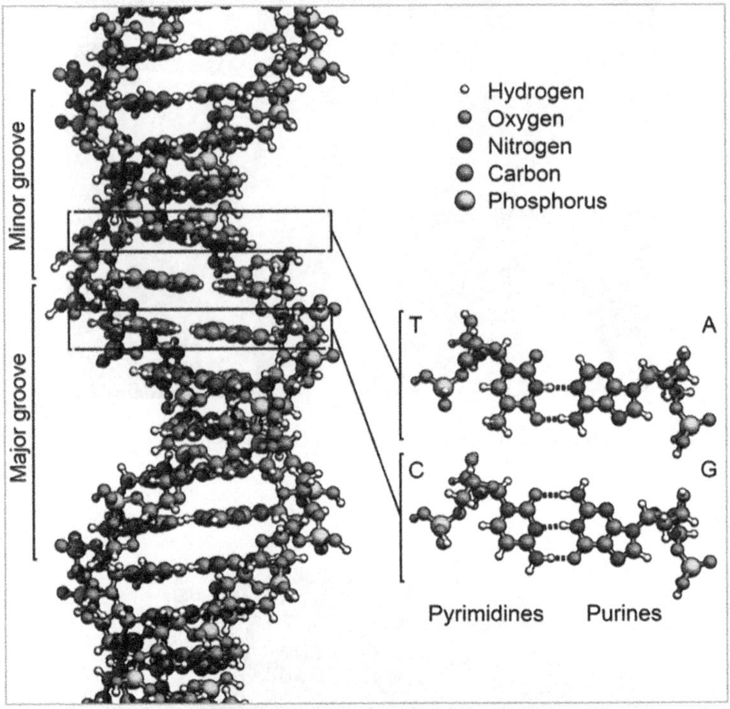

The structure of the DNA double helix. The atoms in the structure are colour-coded by element and the detailed structures of two base pairs are shown in the bottom right.

Fig. 6.16 Courtesy of https://en.wikipedia.org/wiki/DNA

Chapter 7
Spiral and Cyclic Structures in the Abiotic Inorganic Material World

The Scale of Spiralicity and Cyclic Motion in the Universe

Spiral structures such as galaxies, hurricanes, tornados, whirlpools and mollusk shells and cyclic motion such as waves and rotations abound in the physical universe in both abiotic inorganic matter and in the biosphere of living organisms. Examples of wave motion includes, to mention a few, ocean waves, sound waves, electromagnetic waves such as light, X-rays and gamma rays. The cyclic rotations include the orbiting of electrons around the nucleus of the atom; the orbiting of the Earth, the other planets, asteroids and comets around the sun; the orbiting of moons around their planets and the orbiting of stars around the center of their respective spiral galaxies. The range in size of these phenomena vary 140 light years across for our own Milky Way Galaxy to the wavelengths of gamma rays emitted by radioactive atoms that are less than 10^{-11} m. atom. A spiral galaxy can therefore be 10^{+26} times bigger than a gamma ray photon. The range in the time scale of rotations is just as vast. They vary from 225 million years for the sun to rotate around the center of the Milky Way Galaxy, to 1 year for the earth to rotate around the sun and 24 h for it to rotate on its axis to the order of 10^{-16} s for the rotation of an electron around the nucleus of an atom and hence and ratio of the period of the Sun's rotation about the center of the galaxy to the period of the electron's rotation about the nucleus of the hydrogen atom is 225 years/10^{-16} s = $225 \times 365.25 \times 24 \times 60 \times 60/10^{-16}$, which is also approximately 10^{+26} (the same as the ration of the size of a galaxy and a gamma ray photon, which is an unexpected). There is a fantastic variation in both the scale of the size of cyclic phenomena as well as the period of their cycles.

R. K. Logan, I. Pruska-Oldenhof, *A Topology of Mind*, Mathematics in Mind,
https://doi.org/10.1007/978-3-030-96436-8_7

Spiral Galaxies

The Whirlpool Galaxy was the first spiral galaxy to be sighted, which occurred in 1773 by Charles Messier. It is approximately 23 million light years away from Earth (see Fig. 7.1 which also shows the much smaller dwarf galaxy cataloged as NGC 5195, which is the companion galaxy of the Whirlpool Galaxy). Astronomers subsequently learned that are own galaxy the Milky Way Galaxy is also a spiral galaxy (see Fig. 7.2 for an artist's rendition of what our galaxy would look like seen from afar and Fig. 7.3 which represents a side view of the Milky Way Galaxy showing the bulge at its center).

The spiral structure of the Milky Way Galaxy is due to the rapid rate of its rotation about its center, where a gigantic black hole is located that has a mass more than 4.1 million times the mass of the sun. The rate at which the stars on the edge of the galaxy are rotating is so great that these stars should fly off into space away from the galaxy given that the total mass of stars and gases contained in the galaxy is not sufficient to generate a strong enough gravitational to keep these stars within the

Fig. 7.1 The Whirlpool Galaxy and its companion galaxy, NG 5195, that are 23 million light years away from earth. (Image courtesy of Wikimedia Commons and NASA)

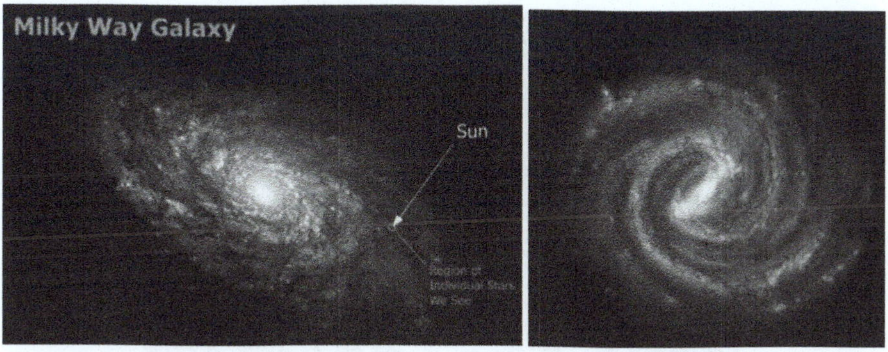

Fig. 7.2 The Milky Way Galaxy. (Courtesy of Wikimedia)

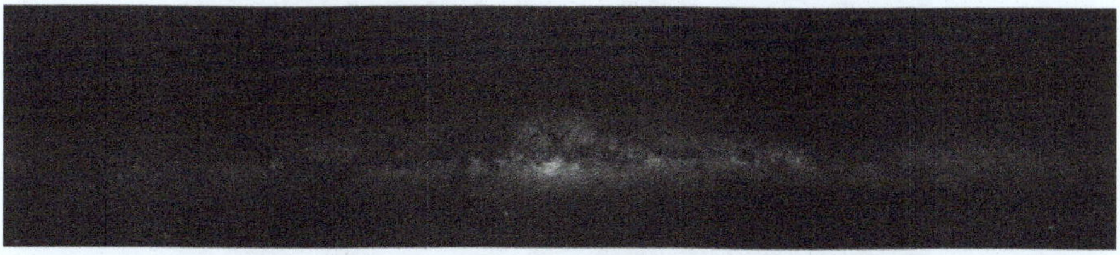

Fig. 7.3 A side view of the Milky Way showing the bulge at its center. (Courtesy of Wikipedia)

galaxy. This is why astronomers and cosmologists have had to postulate the existence of dark matter in order to account for the rapid rotation of the galaxy and the fact that it does not break apart due to the centrifugal force that its rapid rotation creates.

Double Helix Nebulae

The Double Helix Nebulae (see Fig. 7.4), composed of cosmic dust, hydrogen and helium gas, is only 300 light years from the super massive black hole at the center of our Milky Way Galaxy. It has a double helix structure like DNA due to the very strong magnetic fields in the region close to the center of the Milky Way. The nebulae stretches out over a distance of 80 light years and contains many stars.

Cyclic Time and Ancient Astronomy

In Chap. 6 we talked about spirals that represent cyclic behavior in space and those that represent cyclic behavior in time as in human history. We referenced time cycles in regard Vico's notion of corso and ricorso, Hegel's dialectic, and Nietzche's and

Fig. 7.4 The Double Helix Nebulae

Fig. 7.5 Stonehenge. (Courtesy of Wikipedia)

Eliade's idea of the eternal return. The notion of cycles in times has long been a part of human knowledge long before the philosophers we just referenced. The rhythm of the seasons due to the earth's rotation around the sun, the phases of the moon, even the cyclic recurrences of eclipses have been part of human cultures for thousands of millennia. Stonehenge is thought to have been both a place of religious worship and an astronomical observatory that was used to make predictions of solstices and equinoxes and even eclipses by a pre-literate culture living on the Salisbury plain in England dating back to 3000 BC (see Fig. 7.5). One cannot help but be impressed when one realizes that the builders of Stonehenge had determined a 56-year cycle of lunar eclipses.

We also possess records of the observations of eclipses and their cyclic recurrences by the Babylonians and the Chinese date back to 2500 BC. Cuneiform tablets known as Enûma Anu Enlil reveal that the Babylonians were aware of the cyclic nature of astronomical events. Table 63 "lists the first and last visible risings of Venus over a period of about 21 years (en.wikipedia.org/wiki/Babylonian_astronomy_years, accessed May 21, 2017)" and reveals that the movements of Venus are periodic and can be described mathematically.

The oldest known Chinese observatory was discovered in 2005 in Shanxi Province and dates back to the Longshan period of 2300–1900 BC. A circular carved platform with a diameter of 60 meters "was used to locate the rising of the sun at the different periods of the year (http://idp.bl.uk/4DCGI/education/astronomy/history.html, accessed May 21, 2017)." Subsequent observations of Chines astronomers revealed the periodic occurrences of eclipses. They were also aware that the light from the moon and the planets was reflected from the sun.

There were also observations in pre-Columbian Meso-America by Mayan (see Fig. 7.6) and Aztec astronomers who were able to predict lunar and solar eclipses, solstice events, the cycles associated with the rotation of Venus around the sun (Martin 1995). The Mayans and the Aztecs developed precise calendars of celestial events, a practice that pre-dates both of these cultures.

> The Maya didn't invent the calendar, it was used by most cultures in pre-Columbian Central America—including the Maya—from around 2000 BCE to the 16th century. The Mayan civilization developed the calendar further and it's still in use in some Maya communities today (https://www.timeanddate.com/calendar/mayan.html)

The Mayan calendar actually consisted of three calendars: the Haab, a civil calendar based on the 365-day solar year; the Tzolkin or divine calendar of 260-days used to determine the time of religious and ceremonial events; and the Long Count calendar, which keeps track of the time for the cyclic destruction and re-creation of the

Fig. 7.6 Mayan "El Caracol" observatory temple at Chichen Itza, Mexico. (Courtesy of Wikipedia)

Fig. 7.7 Aztec Calendar. (Courtesy of Wikipedia)

world. According to the Mayan Long Count calendar the world in this cycle began on September 6, 3114 BC and was scheduled to end on winter solstice, December 21, 2012. The Aztec calendar had many features of the Mayan calendar namely the 365-day solar year for civic activities and the 260-day year for keeping track of ritual events (see Fig. 7.7).

The Weekly Cycle

All cultures have or have had the notion of a cycle of days known as a week that is considerably less than a month. As its very name suggests the cycle of a month (or moonth) is easy to explain in terms of the monthly rotation of the moon around the Earth. The notion of a week consisting anywhere from 3 to 10 days is harder to explain. Today almost all of the contemporary world operates on a weekly cycle of 7 days, but that has not always been the case. The Achaemenid calendar of ancient Persia had three 7-day weeks every month followed by a fourth and final week of 8 or 9 days depending on the month. This reinforces the notion that the notion of a week might have arisen from the idea of breaking the month up into four quarters. Ancient Rome used an 8-day week as seems to be the case for the pre-Christian Celts. Traces of a 9-day week are found in the languages of the Baltic people and also the Welsh. The ancient Chinese and Egyptian calendars featured a 10-day week. The Akan people of west Africa used a 6-day week. An analysis of the name days in Gipuzkoan dialect hints at the possibility the Basque people once has a 3-day week. The Soviet government in Russia played with the notion of a 5-day week and a 6-day week between 1929 and 1940 as a way to increase production and to discourage the use of Sunday as a day of rest as a way of attacking religion.

What is interesting about the different schemes for the definition of a week is the apparent universality of a short-term cycle of days that goes by the notion of a week. It hints at one of the hypotheses of our study that thinking in terms of cycles is a universal characteristic of the human mind and its mathematical nature.

A Cyclical Model of the Evolution of the Universe

One of the competing cosmological models in physics entertains the notion of cyclic time in terms of the universe as a whole. It is posited that the Big Bang that started the current evolution and expansion of the universe and that that will be followed by an eventual slowing down of the expansion of the universe to the point where it stops expanding and eventually starts to contract resulting in what the theory designates as the Big Crunch. After which another cycle will begin with another Big Bang, another expansion, followed by another contraction and another Big Crunch and so on and so forth ad infinitum. This is only a theory and a speculative one, at that, without much evidence to support it. In fact, contrary to this model the expansion of the universe seems to be accelerating, resulting in many cosmologists accepting the notion that the acceleration of the universe will continue and that is due to what they call dark energy, which they claim is undetectable.

Does history in the cosmic sense repeat itself? Still another possibility is that the universe began an infinite time ago as an infinitesimally dilute gas of infinite extent, which began collapsing. Then 13.7 billion years ago the universe had collapsed down to a point and exploded. The universe is now expanding and will continue to expand forever eventually returning to the state from which it began. In this model there is only one cycle of the universe.

This model and the model with the successive Big Bangs and the Big Crunches just discussed in which the universe is continually oscillating between expanding and shrinking phases, have one feature in common. In both models the universe always was and always will be. We personally find this an appealing principle because of the conservation of energy. If the universe were suddenly to start at time zero from nothing, we would need to understand how such a colossal violation of energy conservation could take place. If the total energy of the universe were decreasing, one could go forward in time to when there would be no energy and hence the end of the universe. If the total energy of the universe were increasing, then one could go backward in time to when there was no energy or the point when the universe had just begun. But since total energy is conserved, it is easy to conceive of a universe that has neither a beginning nor an end. It always was and always will be "an ever-kindling fire" as Heraclitus described it 2500 years ago.

Electromagnetic Waves and Cycles

Gamma rays, the smallest form of cyclic behavior, with a wave length of $\lambda < 10^{-11}$ m, are a form of electromagnetic radiation. Visible light with a wave length of λ of approximately 5×10^{-7} m and radio waves with wavelengths between 10 and 100 m are also forms of electromagnetic radiation, which consists of the oscillations of the electric and magnetic fields which mutually induce each other to oscillate back and forth. Rather than explaining the electric and magnetic forces between individual moving charged particles as Newton did for gravitational forces between masses, Michael Faraday introduced the notion of the electric and magnetic fields. A charged particle creates an electric field and a moving charged particle creates a magnetic field. These two fields then exert forces on other moving or static charged particles. James Clerk Maxwell was able to express all the laws of electricity and magnetism in terms of four very simple equations, which relate the electric and magnetic fields and show their intimate connection. Maxwell showed that his equations could explain how a changing magnetic field creates a changing electric field and how a changing electric field creates a changing magnetic field. His equations explained the propagation of visible light, gamma rays and other forms of electro-magnetic radiation, which are nothing more than oscillating electric and magnetic fields which mutually induce each other's oscillations or spiraling behaviors.

The concept of an electromagnetic wave is easy to understand once its relation to electric and magnetic induction is realized. Consider an electric field oscillating at some point in space. Then, by magnetic induction, (or since a changing electric field produces a magnetic field) the oscillating electric field creates a magnetic field perpendicular to itself in its immediate neighborhood. This oscillating magnetic field, as a result of electric induction, then creates an oscillating electric field, which, in turn, induces an oscillating magnetic field, which, in turn, induces an oscillating electric field and so on and so forth. In this way, an electromagnetic wave propagates through empty space at the velocity of light, denoted by c, which is 300,000 kilometers per second (186,000 miles per second) or 3×10^8 meters per second.

Electromagnetic radiation explains a large range of phenomena which depend on the frequency of the oscillations or cycles, f, and the length of the waves that are propagated, λ. Here is a list of the different forms of radiation and their frequencies, f, and wavelengths, λ. The wavelength and frequency are related by the equation $\lambda = c/f$.

Table of radiation, wavelength and frequency of electromagnetic radiation

Type of wave	Wavelength, λ (cm)	Frequency, f (s^{-1})
Radio waves	10^3 to 10^4	10^8 to 10^9
Microwaves	10^{-3} to 10	10^9 to 10^{12}
Infrared (heat)	10^{-4} to 10^{-3}	10^{13} to 10^{14}
Visible light	$4–7 \times 10^{-5}$	10^{15}
Ultra-violet	10^{-7} to 10^{-6}	10^{16} to 10^{17}
X-rays	10^{-9} to 10^{-7}	10^{17} to 10^{19}
Gamma rays	Less than 10^{-9}	10^{19} and higher (ibid.)

Whirlpools, Maelstroms and Vortices

The flow of fluids whether they are gases or liquids gives rise to spiral motions as illustrated below in Figs. 7.8, 7.9, 7.10, 7.11, 7.12 and 7.13 when two streams of a gas or water flow into each other from opposite directions creating a torque and hence a swirling body of water as is the case with whirlpools, maelstroms and

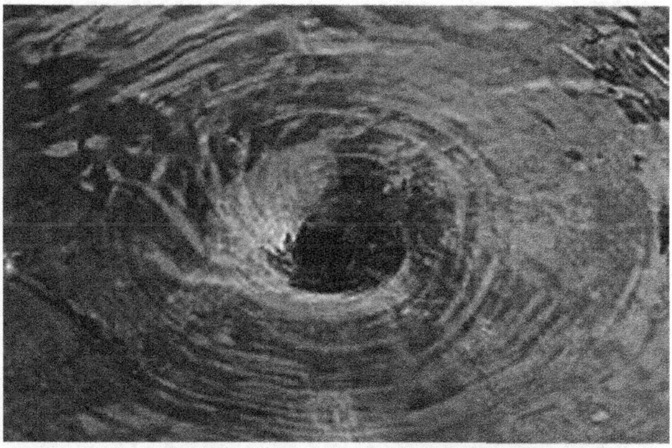

Fig. 7.8 Whirlpool in a pond. (Image courtesy Wikimedia Commons)

Fig. 7.9 Niagara Whirlpool. (Image courtesy Wikimedia Commons)

vortexes and a swirling body of air as is the case with cyclones, hurricanes, and tornados.

Obstructions in a stream of water in a river or a brook can create a back water from which a small whirlpool can arise (see Fig. 7.8). Even weaker whirlpools can also be observed in sinks and bathtubs when they are draining. Whirlpools can also

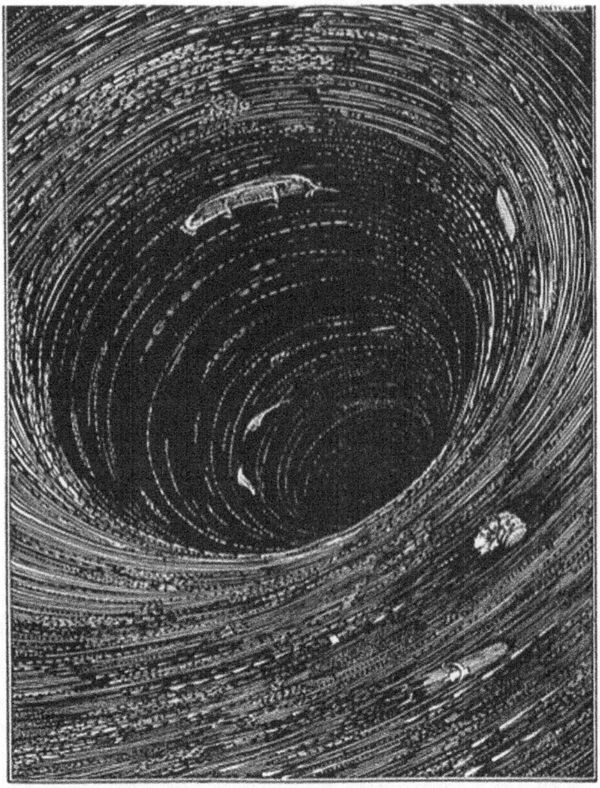

Fig. 7.10 Artist Harry Clarke's 1919 illustration for "A Descent into the Maelström"

form at the base of a waterfall. A famous one is the Niagara Whirlpool that is 35 m deep located in a basin formed by steep rocks cliffs that the flow of the Niagara River carved out over the millennia somewhat downstream from Niagara Falls (see Fig. 7.9).

Maelstroms, often caused by the actions of tides flowing in narrow straits, are powerful whirlpools that give rise to a downdraft that have on some occasions caught small ships and destroyed them. Saltstraumen, the strongest maelström in the world, is located off the north coast of Norway near the Arctic Circle. Corryvreckan is a narrow strait between the islands of Jura and Scarba, on the northern side of the Gulf of Corryvreckan on the north Scottish coast. It gives rise to the third-largest whirlpool or maelström in the world. The most famous maelström of all is Moskstraumen off the north coast of Norway, made famous by Edgar Allen Poe's 1841 short story "A Descent into the Maelström." In the tale, a mariner recounts how he survived a shipwreck in the Moskstraumen Maelström by grabbing hold of a barrel that rose to the top of the whirlpool after observing that wooden object like pieces of wood were not dragged down by the downdraft but rather rose

Fig. 7.11 A clockwise spinning extratropical cyclone off of southern Australia. (Courtesy Wikimedia Commons)

Fig. 7.12 2005 Hurricane Katrina. (Courtesy Wikimedia Commons)

to the top of the whirlpool (see Fig. 7.10). Marshall McLuhan used this story to underscore the importance of observation.

> Poe imagines the situation in which a sailor, who has gone out on a fishing expedition, finds himself caught in a huge maelstrom or whirlpool. He sees that his boat will be sucked down into this thing. Pattern recognition in the midst of a huge, overwhelming, destructive force is the way out of the maelstrom. The huge vortices of energy created by our media present us with similar possibilities of evasion or consequences of destruction. By studying the patterns of the effects of this huge vortex of energy in which we are involved, it may be possible to program a strategy of evasion and survival (http://marshallmcluhanspeaks.com/lecture/1979-man-and-media, accessed May 20, 2017).

Fig. 7.13 Tornado in Elie Manitoba. (Image courtesy Wikimedia Commons)

Cyclones, Hurricanes, Tornados

Just as whirlpools form in water whereas opposing streams create a torque the same thing happens in the atmosphere with opposing currents of air. Weather systems on the scale of hundreds of miles give rise to cyclones in the south Pacific Ocean, typhoons in the northwest Pacific Ocean and hurricanes in the Atlantic Ocean (see Figs. 7.11 and 7.12). These storms are characterized by a rapidly rotating storm system, and a spiral arrangement of thunderstorms, lots of rain and a low-pressure area in the eye of the storm which is most often quite calm. The eye wall surrounding the eye is where the winds are the strongest.

Tornados (see Fig. 7.13) are rapidly rotating column of air in the shape of a funnel that originate at the base of a cumulonimbus or a cumulus cloud and create great damage when they touch down on surface of the earth. Tornados occur when the warm air in the updraft of a storm system collides with the cold air in the storm's downdraft causing a rotation. The size of a tornado is much smaller than a hurricane and have a diameter in the range of 0.1–3 km.

The Cycle of Ice Ages

The surface of the Earth and its atmosphere is constantly warmed by sunlight from the sun. A certain amount of the heat from the atmosphere and the surface of the Earth is radiated back into outer space preventing the planet from overheating to the point that life on our planet would not be sustainable. The delicate balance in nature in which the amount of incoming energy and outgoing energy are approximately equal so that the temperature on Earth has remained in the moderate range. Volcanic activity, fluctuations in the sun's output and fluctuations in the Earth's orbit about the sun have given rise to cycles of ice ages and to moderate climates known as Milankovich cycles. The last Ice Age ended 10,000 years ago. The fluctuations in albedo, the reflection of sunlight back into outer space due to ice and snow cover has accelerated the transitions from and to Ice Ages. Another factor affecting the amount of heat loss due to radiation of energy into space is the amount of carbon dioxide, CO_2, in the atmosphere, which is approximately 0.0383% of all the gases in the atmosphere. Because CO_2 absorbs a significant percentage of the heat radiated off the surface of the Earth, its presence in the atmosphere contributes to the warming of the planet. This is the problem we face today with climate change and global warming.

Sunspot, Solar Flare Cycles and the Van Allen Radiation Belt

The surface of the sun is not uniform but mottled. It is peppered with granules, a 1000 km in diameter, whose centers are hotter than their edges due to hot gases rising to the surface of the sun. There are also sunspots caused by disturbances of the sun's magnetic field and may be regarded as magnetic storms of the sun. Sunspots are dark because they are cooler than the surrounding surface. Sunspot activity is cyclic and recurs approximately every 11 years during which solar prominences and solar flares frequently take place. Solar prominences shoot out tremendous jets of solar material several thousand kilometers above the surface of the sun after which they fall back again into the sun. Solar flares are sudden outbursts of radioactive energy which give rise to solar winds of energetic charged particle.

The Earth is surrounded by the Van Allen radiation belt, which is a plasmatic torus made up of energetic charged particles trapped by Earth's magnetic field. The charged particles making up the Van Allen belt arrive on the wings of the solar wind. The Earth's geomagnetic field varies by ½ of 1% due to the effects of the Sun, the solar wind, and the Van Allen radiation belts. There is also a variation due to the Sun spot activity that comes in an 11-year cycle.

Miscellaneous Spiral Structures

Another place of spiral activity in our solar system is in the rings of Saturn (see Fig. 7.14), which contain a great many spiral density waves due to the perturbation caused by Saturn's inner moon.

Another example of a spiral structure is the planetary nebula, NGC 5189 (see Fig. 7.15). It formed around a star in a binary star system 1780 light years away. It is not really a planet but rather a glowing shell of ionized gas.

In 2007 photos were made of the Comet McNaught (see Fig. 7.16). "The images reveal three jets of gas that spiral away from the comet's rotating nucleus (http://www.space.com/3505-photos-reveal-great-comet-spiraling-jets.html, accessed May 21, 2017)."

Another example of a spiral structure on the surface of the earth is the Richtak Structure (http://www.physics-astronomy.com/2014/10/this-huge-spiral-structure-in-sahara-is.html#.WSISkhgZPfY) located in Mauritania in Africa. It is visible

Fig. 7.14 Saturn's Rings. (Courtesy of Wikipedia)

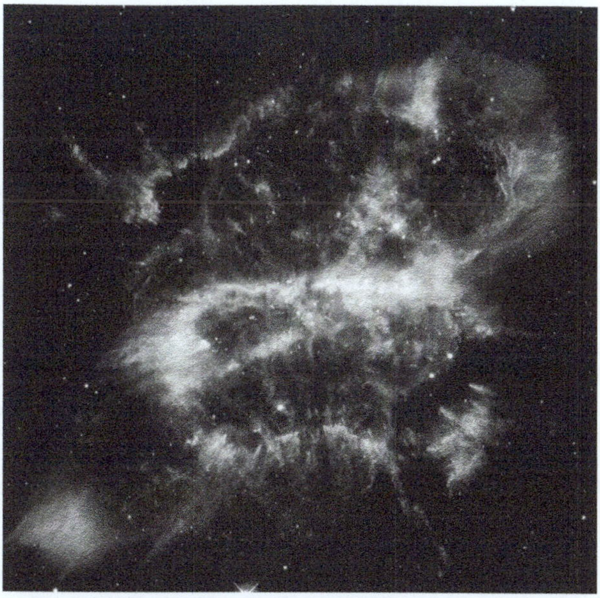

Fig. 7.15 NGC 5189 Planetary Nebula. (Courtesy of Wikipedia)

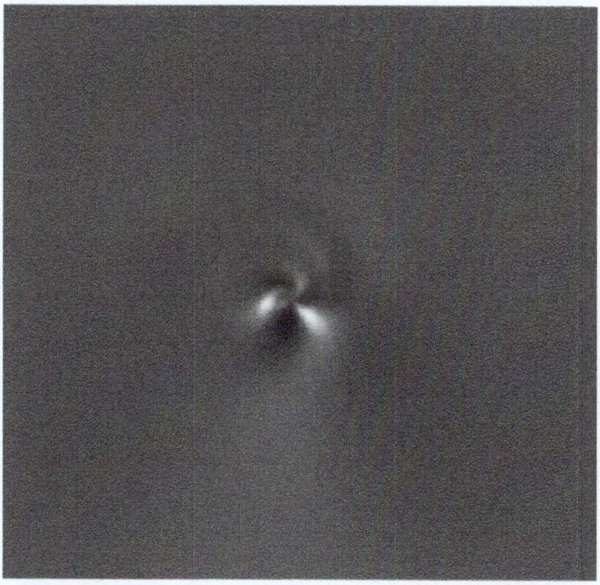

Fig. 7.16 Richtak Structure

Fig. 7.17 Richtak Structure in Mauritania. (Courtesy of Wikipedia and www.physics-astronomy.com respectively)

from the International Space Station (see Fig. 7.17). The geological formation is the outcome of slow erosion of the soil.

Conclusion

The physical abiotic universe at various scales from the size of galaxies to gamma rays, the electro-magnetic radiation with the shortest wave lengths, is full of spiral structures. We next turn to the spiral structures in living organisms in the next chapter.

Chapter 8
Life Is a Spiral and It Is Cyclic: The Spiral Structures of the Biosphere

The Seven Ages of Man
All the world's a stage,And all the men and women merely players,
They have their exits and entrances,And one man in his time plays many parts,His acts being seven ages.
At first the infant, Mewling and puking in the nurse's arms…
Last scene of all, That ends this strange eventful history,Is second childishness and mere oblivion,Sans teeth, sans eyes, sans taste, sans everything.

William Shakespeare

Introduction: Life Began on Planet Earth with the Spirals of RNA and DNA

It is believed by most biologists that life began on this planet with the formation of ribonucleic acid (RNA) illustrated in Fig. 8.1 that have the property of being self-replicating as well as the ability to catalyze the production of the chemical groups and information transfers required for life within living cells. The structure of RNA is that of a double helix like that also of deoxyribonucleic acid (DNA) illustrated in Fig. 8.2. RNA consists of a single chain of nucleic acid that doubles back on its self to form a double helix. DNA, on the other hand, is a double helix consisting of two separate strands of nucleic acid that link up with each other. Although it is thought that life began with RNA it eventually was replaced with the current RNA-DNA-protein regime that characterizes life today. The role of DNA is the storage of information and its role in catalyzing the production of RNA. The role of RNA is that of catalyzing the production of proteins which actually carry out the metabolic functions of day to day life of a cell. Many proteins also have spiral components such as the α-helix illustrated in Fig. 8.3, one of the most prevalent of the local structures in proteins. Biochemical spirals are an essential element of living organisms without which there would be no life.

R. K. Logan, I. Pruska-Oldenhof, *A Topology of Mind*, Mathematics in Mind,
https://doi.org/10.1007/978-3-030-96436-8_8

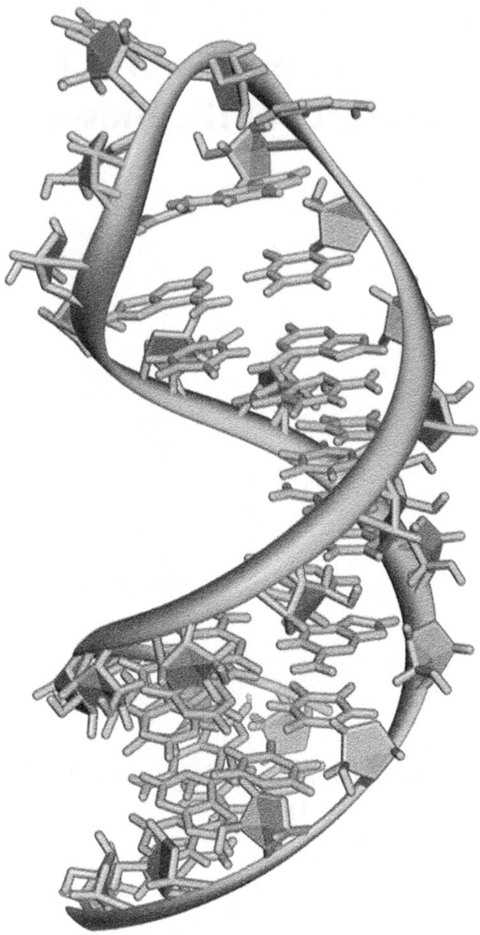

Fig. 8.1 RNA. (Courtesy of Wikipedia)

Extra-Cellular Spiral Structures in Animals and Plants

Collagen

In addition to the spiral structure of the biochemical molecules of DNA, RNA and proteins within the structure of living cells spiral structure abound in the collagen, the main structural protein in the extracellular space in the various connective tissues in animal bodies. Collagen is found in bones, muscles, tendons, ligaments, cartilage, skin, corneas, blood vessels, the guts, intervertebral discs, and the dentine in teeth. The secondary structure of collagen consists of a triple helix made of repetitive amino acid sequences as shown in Fig. 8.4 of the collagen triple helix.

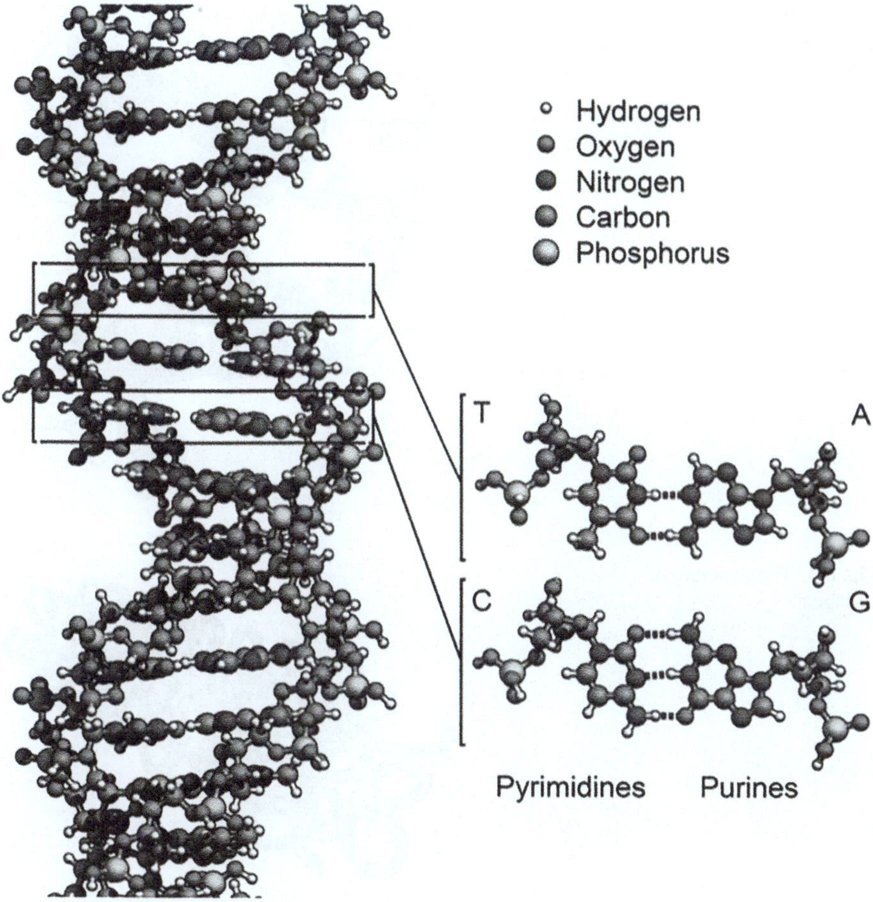

Fig. 8.2 DNA. (Courtesy of Wikipedia)

Human Stomach Muscles

The arrangement of stomach muscles that prevents food from traveling backward consists of an inner circular layer with a steep pitch helical structure and a longitudinal outer layer also with a helical structure, but one that is shallower.

Spiral Structure in Plants and Animals

The spiral structure can be found in the biosphere of living organisms such as is the case with sunflower florets (Fig. 8.5), the nautilus mollusk shell (Fig. 8.6), Romanesco broccoli (Fig. 8.7), the tail of the veiled chameleon (Fig. 8.8), the

Fig. 8.3 A protein with α-helices. (Courtesy of Wikipedia)

Fig. 8.4 The structure of collagen

seahorse (Fig. 8.9), the leaves of the Aloe polyphylla (Fig. 8.10), the cone of the Queen Sago cycad (Fig. 8.11), the horns of the Rocky Mountain big horn sheep (Fig. 8.12), Spiraling shell of the snail Trochoidea liebetruti (Fig. 8.13) and the shape of a slithering snake (Fig. 8.14).

We begin the gallery of spirals in various biological specimens by reviewing the three following examples: the florets of the sunflower, which has the pattern of the Fermat's spiral that allows the maximum number of seeds that can be packed into the flower; the logarithmic spiral of the Nautilus Mollusk Shell; and the fractal spiral structure of Romanesco broccoli. We continue thereafter with examples from the world of plants and animals to catalog some of the spiral and cyclic structures of living organisms.

Fig. 8.5 Sunflower florets

Fig. 8.6 Nautilus mollusk shell

Aloe Polyphylla

The leaves of the aloe polyphylla grow in a spiral shape that can be either clockwise or counter clockwise. The leaves are staggered in this spiral pattern to permit optimum exposure to sunlight.

Fig. 8.7 Romanesco broccoli

Fig. 8.8 The tail of the veiled chameleon

Fig. 8.9 A seahorse

Fig. 8.10 Aloe polyphylla. (Image courtesy of the Wikimedia Commons)

Human Cochlea of the Inner Ear

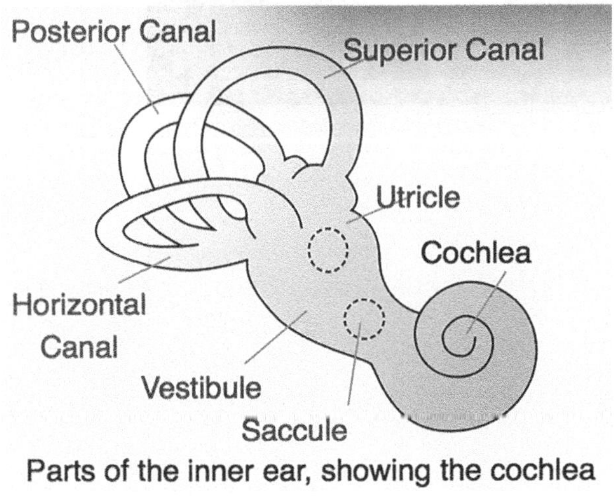

Courtesy of Wikimedia

Fig. 8.11 Cone of Queen Sago cycad. (en.wikipedia.org/wiki/Patterns_in_nature, Courtesy of Wikipedia)

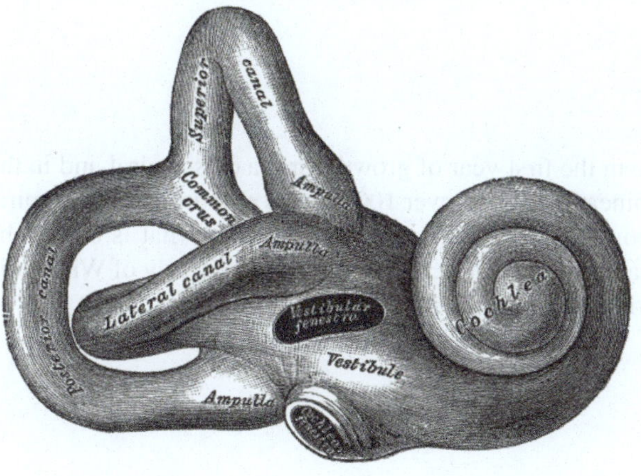

Courtesy of Wikimedia

Fig. 8.12 Horn of Rocky Mountain sheep. (en.wikipedia.org/wiki/Patterns_in_nature, Courtesy of Wikipedia)

The cochlea, the auditory portion of the inner ear is a spiral-shaped cavity in the bony labyrinth, in humans making 2.5 turns around its axis.

Pineapples

The leaves of the pineapple in the first year of growth form a close spiral and in the second year of growth the pineapple forms over 100 flowers also arranged in a spiral formation, which then becomes berries that coalesce and form what is called the pineapple which has also has spiral like structure to it. Image courtesy of Wikipedia (https://en.wikipedia.org/wiki/Pineapple).

Fig. 8.13 Spiraling shell of the snail Trochoidea liebetruti. (en.wikipedia.org/wiki/Patterns_in_nature, Courtesy of Wikipedia)

Fig. 8.14 The meandering of a sinuous snake crawling. (Courtesy of Wikimedia)

Pineapple and its cross-section

A young pineapple in flower

In Chap. 6 we noted the relationship of Fibonacci numbers 1, 1, 2, 3, 5, 8, 13, 21, 34 to the structure of a spiral. The number of flower petals for the most part is one of these Fibonacci numbers. Here are some examples collected on the Web page http://britton.disted.camosun.bc.ca/fibslide/jbfibslide.htm.

White Cala Lily—1-petal

Euphorbia—2-petals

Trillium—3-petals

Columbine—5-petals

Bloodroot—8-petals

Black-eyed Susan—13-petals

Shasta Daisy—21-petals

Field Daisy—34-petals

Multi-petal Flower—55-petals

The female pine that produces the seeds of a pine tree have scales that are overlapping like the scales on a fish. The scales are arranged in a spiral pattern in the ratios of the Fibonacci numbers just like the number of petals on flowers.

Coulter Pine Cone

Colorado Blue Spruce Cones

Chapter 9
Spiral Thought Structures in History and Philosophy

Introduction

The cyclic/spiral structure of human thought patterns is a universal characteristic of many cultures throughout human history. The idea of eternal return is the notion that events in the universe and in particular the events experienced by humankind repeat themselves so that they occur and re-occur for ever and ever. This notion dates back to ancient Indian and ancient Egyptian philosophy. The ancient Egyptians also believed in the notion of the eternal recurrence with their notion of neheh, roughly translated as endless recurrence. It was also an idea at the core of the beliefs of the Pythagoreans and the Stoics.

Throughout history spiral thought structures have characterized accounts of human history, philosophy and psychology ranging from the most ancient cultures to the most recent thinkers of the nineteenth, twentieth, and twenty-first centuries. To give the reader an idea of the scope of this phenomenon the following topics will be covered in this chapter (they are listed in a more or less chronological order):

- Ancient Indian notions of cyclic time (3300 BC to present)
- Ancient Chinese notion of the dynastic cycle (2070 BC to the present)
- Ancient Greek political cycle of Kyklos and Pythagorean recurrence (500 BC)
- Ancient Hebrew Biblical Idea of Recurrence (453 BC)
- Arab historian Ibn Khaldūn (1377)
- Vico's notion of corso and ricorso (1725)
- Hegel's dialectic philosophy (1807)
- Schopenhauer (1818)
- Nietzsche's Eternal Return (1882)
- Camus' Myth of Sisyphus (1942)
- Eliade's Eternal Return (1971)

R. K. Logan, I. Pruska-Oldenhof, *A Topology of Mind*, Mathematics in Mind,
https://doi.org/10.1007/978-3-030-96436-8_9

This list is by no means complete but it will give the reader a sense of cyclic thinking in the fields of history, philosophy and psychology through the centuries.

Ancient Indian Notions of Cyclic Time (3300 BC)

There are a variety of religions practised today in India by Hindus, Buddhists, Jains and Sikhs with their roots in ancient Indian religions. Each of religions embrace some form of cyclic time, known as saṃsāra, the idea of birth, death and rebirth or reincarnation as well as moksha or the release from the cycle of saṃsāra. According to Hindu teaching there are four ages or yugas. The first age of Satya Yuga is the age of truth when humankind acted in accordance with the truth as recorded in the Vedas. It was followed by the Trata Yuga in which one quarter of the truth was lost and then the Dwapar Yuga in which half the truth was lost leaving humanity in the current Kali Yuga with only one quarter of the truth remaining in which humankind lives primarily in sin. The Kali Yuga will eventually end when the universe will be destroyed by Lord Shiva and the universe will be re-created by Lord Brahma and the four yugas or ages will begin once again.

The Ancient Chinese Notion of the Dynastic Cycle (2070 BC)

Chinese historians as far back as 2070 BC formulated the notion of dynastic cycles in which a new dynasty emerges replacing an older one. At first the new dynasty is vital and dynamic but over time it degenerates until it is replaced by a new vital dynasty and the eternal cycle resumes once again.

China has been ruled by a succession of emperors belonging to a succession of dynasties each of which was said to have possessed what was called the Mandate of Heaven. According to the dynastic cycle each new dynasty that followed the older dynasty it replaced began with political, economic and cultural success in which China flourished. Over time, however, a decline in the well-being of state occurred as a result of the abuse of power and general corruption of those in power. As a result of this decline the corrupted dynasty was replaced by a new dynasty and once again the cycle of flourishing and prosperity followed by abuse of power, decline and replacement by a new dynasty. Not all dynasty consisted of Han Chinese but there were conquerors from time to time. This cycle has continued to the present day with the succession of President *Chiang Kai-shek* followed by Maoism followed by today's state form of capitalism.

Ancient Hebrew Biblical Idea of Recurrence (453 BC)

The idea of recurrence can be found in Hebrew scriptures in the much-quoted passages from Ecclesiastes:

> What has been will be again, what has been done will be done again; there is nothing new under the sun (Ecc 1:9).
>
> Is there anything of which one can say, "Look! This is something new"? It was here already, long ago; it was here before our time (Ecc 1:10).

A similar theme of recurring cycles can be found in Genesis:

> While the earth remains, seedtime and harvest, cold and heat, winter and summer, and day and night shall not cease (Gen 8:22).

Another form of recurrence in the Jewish tradition was the cyclic notion of the 7-day week and the notion of a sabbath on the seventh day. This recurring pattern of time was formulated for the first time in human history by the Jews during their period of exile in Babylon following the destruction of the first Temple (Zerubavel 1989). The mention of the 7-day cycle or the cycle of the week occurs in Hebrew scriptures in the following passages:

> Gen 1:31 And God saw everything that He had made, and, behold, it was very good. And there was evening and there was morning, the sixth day.
>
> Gen 2:1 And the heaven and the earth were finished, and all the host of them.
>
> Gen 2:2 And on the seventh day God finished His work which He had made; and He rested on the seventh day from all His work which He had made.
>
> Gen 2:3 And God blessed the seventh day and hallowed it; because that in it He rested from all His work which God in creating had made.

The idea that humankind should also rest one day of the week is also found for the first time in Hebrew Scriptures in the book of Exodus and that day is the last day of the week, the Sabbath:

> Ex 20:7 Remember the sabbath day, to keep it holy.
>
> Ex 20:8 Six days shalt thou labour, and do all thy work;
>
> Ex 20:9 But the seventh day is a sabbath unto the LORD thy God, in it, thou shalt not do any manner of work, thou, nor thy son, nor thy daughter, nor thy man-servant, nor thy maid-servant, nor thy cattle, nor thy stranger that is within thy gates;
>
> Ex 20:10 For in six days the LORD made heaven and earth, the sea, and all that in them is, and rested on the seventh day; wherefore the LORD blessed the sabbath day and hallowed it.

Ancient Greek Myths, the Political Cycle of Kyklos and Pythagorean Recurrence (500 BC)

There are two Greek myths that incorporate the notion of a recurring cycle, namely the myth of Sisyphus and the myth of Prometheus. According to legend Sisyphus was able to put Death in chains so that mortals would not have to die. Death was able to escape, however, and together with the gods condemned Sisyphus to the endless cycle of pushing a heavy rock up a mountain only to have it roll back down the mountain and have Sisyphus return to the bottom of the mountain and once again push the rock to the top of the mountain and so on and so forth for eternity. The myth of Prometheus whose name literally means forethought is another story of the punishment of a mortal. Prometheus is credited with making mortals out of clay and then defying the gods he stole fire from them and gave it to humankind. Zeus was enraged by this act of theft and condemned Prometheus to be tied to a rock and have an eagle, a symbol of Zeus, come every morning and feed on his liver. Prometheus' liver would grow back over night and the eagle would come again and feast on Prometheus's liver.

Kyklos is a traditional cyclic model of how political regimes evolve that was also described in Plato's Republic. The idea was later enlarged upon and elaborated by the Greek historian Polybius (200–118 BC). The cycle begins with society in anarchy or lack of government from which a strong figure emerges as a monarch. At first this form of government works well but as a result of the inheritance of monarchs lacking moral stature, the monarchy degenerates into a form of tyranny. The tyrannical regime is overthrown by the prominent citizens or aristocrats of the state to form an oligarchy that rules effectively until it degenerates through the corruption of the inheritors of the oligarchy. This oligarchy is overthrown by the ordinary citizens of the state to form a democracy which in turn degenerates into rule by the mob or anarchy and the cycle begins all over again.

Another form of cyclic thinking was developed by Pythagoras and his followers. They believed that everything progressed in predictable cycles, which might have been motivated by their knowledge of the cyclic movement of the heavenly bodies including the annual movement of the sun, the monthly movement of the moon as well as the recurring cycles of the planets.

Porphyry, a philosopher from Tyre in the Roman empire wrote in *The Life of Pythagoras* that:

> The following became universally known: first, that he [Pythagoras] maintains that the soul is immortal; second, that it changes into other kinds of living things; third, **that events recur in certain cycles and that nothing is ever absolutely new**; and fourth, that all living things should to be regarded as akin. Pythagoras seems to have been the first to bring these beliefs into Greece (the bolding is ours) (Hakim 2005, 20).

Arab Historian Ibn Khaldūn (1377)

Muhammad ibn Khaldūn al-Ḥaḍramī was a Tunisian/Arab historian who developed a cyclic theory of the rise and fall of empires in which an empire is formed, prospers and after a period of decline is conquered by another regime which creates its empire but incorporates some of the cultural elements of the conquered empire. The conquering empire then suffers the same fate as the empire it conquered and so on and so forth in a never-ending cycle.

Vico's Notion of Corso and Ricorso (1725)

Giambattista Vico's philosophy of history involved the notion of an advancement, a corso followed by a *ricorso* or return. According to Vico, a society evolves to a high point in its development, its *corso*, and then regresses back to or returns to a more primitive time in its history, which he refers to as a *ricorso*. After the *ricorso* the society once again embarks on a new *corso* and progresses to a more advanced level of development to once again experience a *ricorso* or return. The spiral structure of history of *corso* followed by *ricorso* followed by another cycle of *corso* and *ricorso* and so on and so forth is characteristic of Vico philosophy of history. For Vico history has a spiral structure of recurring cycles of development and collapse. The cycle consists of three ages. The first and most primitive is the age of gods. The second age is the age of heroes, in which there is constant conflict between the rulers and the governed. The third age is the age of the people in which democracy emerges but which eventually collapses because of corruption and returns to the age of gods once again. There is some parallel with the Greek notion of Kyklos if we take the age of gods as monarchy given some monarchs or emperors claimed divinity or claimed to have been chosen by God. The age of heroes would correspond to the regime of the aristocrats.

Vico's work was largely ignored by his contemporaries. His influence was not felt until the nineteenth century when his ideas influenced Marx, Goethe, Humboldt, Dilthey, Nietzsche and Gedamer and into the twentieth century when his influence was felt by James Joyce, Marshall McLuhan and Mircea Eliade.

The Spiral Structure of Dialectical Thinking of Fichte and Hegel

The structure of dialectical thinking consists of a triad of elements that include a thesis, its antithesis and the resolution of the thesis and the antithesis through the process of synthesis. The newly arrived synthesis becomes a new thesis which gives rise to a new antithesis which is resolved with a new synthesis and so on and so forth in a spiral of new thinking. The triad of thesis, antithesis and synthesis often attributed to Hegel was actually first formulated by Johann Gottlieb Fichte as a way of describing change. In his *Grundriss des Eigentümlichen der Wissenschaftslehre, in Rücksicht auf das theoretische Vermögen* (The Theoretic Foundation of the Distinctive Character of Scientific Thinking) he wrote, "Die jetzt aufgezeigte Handlung ist thetisch, antithetisch und synthetisch zugleich," which is roughly translated as "The action which is now described is simultaneously thetic, antithetic, and synthetic." Here for the first time the notion of thesis, antithesis and synthesis is articulated.

The foundation of Hegel's philosophical thinking was also formulated in terms of a triad similar to thesis, antithesis and synthesis, namely being, nothingness and becoming and sometimes immediate, mediate and concrete. In Hegel's system Being gives rise to its opposite non-being, which is Nothingness and then by aufheben (sublation) transits into Becoming and Becoming results in Being giving rise to Nothingness and sublating into Becoming and so on ad infinitum. It is a process of recurrence, a never-ending spiral structure of being, non-being or nothingness giving rise to becoming.

Nietzsche's Notion of Eternal Return (1882)

Friedrich Nietzsche first formulated his notion of 'eternal return' in Aphorism 341, entitled 'The greatest weight,' in Book IV at the very end of Die Fröhliche Wissenschaft *(The Gay Science)* where he wrote:

> What, if some day or night a demon were to steal after you into your loneliest loneliness and say to you: "This life as you now live it and have lived it, you will have to live once more and innumerable times more; and there will be nothing new in it, but every pain and every joy and every thought and sigh and everything unutterably small or great in your life will have to return to you, all in the same succession and sequence—even this spider and this moonlight between the trees, and even this moment and I myself. The eternal hourglass of existence is turned upside down again and again, and you with it, speck of dust!"
>
> Would you not throw yourself down and gnash your teeth and curse the demon who spoke thus? Or have you once experienced a tremendous moment when you would have answered him: "You are a god and never have I heard anything more divine." If this thought gained possession of you, it would change you as you are or perhaps crush you. The question in each and every thing, "Do you desire this once more and innumerable times more?" would lie upon your actions as the greatest weight. Or how well disposed would you have to become to yourself and to life *to crave nothing more fervently* than this ultimate eternal confirmation and seal?

The idea of eternal return was not for Nietzsche just a poetic or philosophical idea but something he believed actually and literally happens, is happening and has happened. He even argued for the idea making use of physics and probability arguing that if there is a finite amount of matter in the universe and there is an infinite amount of time then every configuration of matter in the universe would have to repeat itself. He even considered studying physics to confirm his hunch of the existence of eternal return. There was merit in his logic as quantum mechanics and the Heisenberg Uncertainty Principle had not yet been discovered. We can only speculate how he would have rethought his position on the eternal return if he had learned about the Uncertainty Principle and the inability to know precisely both the position and momentum of a particle.

Nietzsche returned to the 'eternal return' in his book, *Thus Spake Zarathustra*, the next book he wrote after the *Gay Science*. In Chapter LVII The Convalescent he explicitly describes his notion of the 'eternal return':

> O Zarathustra, who you are and must become: behold, you are the teacher of the eternal return, that is now your fate! That you must be the first to teach this teaching—how could this great fate not be your greatest danger and infirmity! Behold, we know what you teach: that all things eternally return, and ourselves with them, and that we have already existed times without number, and all things with us. You teach that there is a great year of Becoming, a prodigy of a great year; it must, like a sand-glass, ever turn up anew, that it may anew run down and run out: So that all those years are like one another in the greatest and also in the small, so that we ourselves, in every great year, are like ourselves in the greatest and also in the small… Souls are as mortal as bodies. But the plexus of causes returns in which I am intertwined, it will again create me! I myself pertain to the causes of the eternal return. I come again with this sun, with this earth, with this eagle, with this serpent—not to a new life, or a better life, or a similar life: I come again eternally to this identical and selfsame life, in its greatest and its small, to teach again the eternal return of all things, To speak again the word of the great noontide of earth and man, to announce again to man the Superman.

Eliade's Notion of Eternal Return (1971)

Mircea Eliade's notion of the "Eternal Return" that all religious practitioners not only celebrate the sacred but are actually participating in ceremonies that re-enact and relive those sacred events that gave rise to the notion of the sacred. This idea is indicative of Eliade's spiral thought patterns. Given the universality of this pattern of "eternal return" among religious practitioners across the globe, Eliade's observation hints at the idea that spiral thought patterns are a universal characteristic of the human mind. Eliade hints that the "eternal return" is not just restricted to religious practitioners when he suggests that secularists scientists also entertain a notion of the "eternal return" to the sacred as they try to understand the origin of the universe and the rules that govern it.

The universal celebration of the New Year in all of the world's religions is a reliving of the sacred act of the creation of the universe by the divine. The timing of the New Year is that of a rebirth. The most common time for the beginning of a new year is in the spring time when life springs back into bloom and when new-borns emerge into the world. For example, the rutting season for many mammals occurs so that the young will emerge in the spring time when the chances for their survival are at a maximum. New Year's Day in the Spring were celebrated (or are still celebrated) in Babylon, Assyria, Israel, Iran, Armenia, India, Kashmir, Nepal, Punjab (among the Sikhs), Sri Lanka, Cambodia, Burma, Laos, Thailand, Bali, and China.

The other timing of New Year celebrations is around the time of the winter solstice, which signals the return or rebirth of the sun. The days grow shorter and shorter during the Fall leading up to the winter solstice suggesting the death of the life-giving sun. The New Year's celebration at this point in time symbolizes a rebirth of the sun and a rebirth of light. This is why New Year's celebrations are marked with lights from oil lamps, candles, as is the case for Hannukah, and fires such as the burning of the yule log at Christmas time, a tradition that originated with Germanic pagan worshippers of the sun.

Eliade argues that within the oral tradition time is not a linear progression of events but rather a cyclic repetition of sacred events that are re-experienced through myths and the performance of rituals, that give meaning to the lives of pre-literate humans. For them there is no distinction between the sacred and the profane or the secular. All of their activities such as hunting, gathering, mating, story-telling, dancing, music and socializing are sacred.

> To summarize, we might say that the archaic world knows nothing of "profane" activities: every act which has a definite meaning—hunting, fishing, agriculture; games, conflicts, sexuality,—in some way participates in the sacred (Eliade 1964, 27–28).

There is no religion within the oral tradition because all activity is invested with the notion of the sacred and hence there is no need for religion. There are violations of the sacred but these are treated harshly by the community and often result in banishment. In oral culture one cannot live with a double standard as is the case in literate societies where people behave in one way on their day of prayer and another way for the rest of the week.

The division between the sacred and the profane only arose within literate cultures as did the notion of religion and the linear notion of time as described by Eliade (1964, vii–viii).

> The essential theme of my investigation bears on the image of himself formed by the man of the archaic societies and on the place that he assumes in the Cosmos. The chief difference between the man of the archaic and traditional societies and the man of the modern societies with their strong imprint of Judaeo-Christianity lies in the fact that the former feels himself indissolubly connected with the Cosmos and the cosmic rhythms, whereas the latter insists that he is connected only with History. Of course, for the man of the archaic societies, the Cosmos too has a "history," if only because it is the creation of the gods and is held to have been organized by supernatural beings or mythical heroes. But this "history" of the Cosmos and of human society is a "sacred history," preserved and transmitted through myths. More than that, it is a "history" than can be repeated indefinitely, in the sense that the myths serve as models for ceremonies that periodically re-actualize the tremendous events that occurred at the beginning of time.

The break boundary between the oral tradition and literate society also represents a schism between the purely sacred characteristic of traditional society and the emergence of the division in literate societies between the sacred and the profane, i.e. the division between religious life or practice and the secular activity of commercial and everyday life.

The roots of today's religions in literate societies can be traced back to the oral tradition. Even the content of the sacred scriptures of today's religion are based on the oral tradition and the pronouncements of their prophets that have been transcribed. In other words, the thought patterns of traditional pre-literate societies were cyclic or spiral; corso and ricorso as Vico suggested. This is also true of the religious practitioners of the post-oral or literate era, whose religions practices and sentiments emerged directly from the oral tradition as we will now demonstrate. We begin with the monotheistic Abrahamic religions of Judaism, Christianity and Islam.

The roots of Judaism are in the oral tradition. The Torah of the Jews was initially memorized. It was not written down until after the destruction of the first Temple. It was during the Babylonian exile of the Israelites that Ezra, the Scribe, committed the Torah to its present day written form. The New Testament of the Christians, the gospels, are based on the sayings of Jesus and not on anything he wrote. In a similar way the Koran is believed to have been orally revealed by Allah to Muhammad at various times during his lifetime through the angel Gabriel (Jibril). It was then memorized and recited to his companions who compiled the written form of the Koran.

A similar pattern exists for the Asiatic religions whose roots also stretch back to the oral tradition. The Samhita the oldest part of the Hindu Vedas follows a similar pattern. It was composed and transmitted orally 3000–4000 years ago whereas its written form only dates to 100 BC. Buddhism following the teachings of the Buddha was transmitted orally and as a result bifurcated into many different forms of the religion as it spread from India throughout the Eastern world. The Buddhist texts were composed my monks transcribing from their memory the words of their Buddhist spiritual leaders.

The sacred book of Zoroastrianism, the Avesta, was composed by Zoroaster or Zarathustra (his Persian name) but the religion itself was by that time an old religion with its roots extending back hundreds of years before to the oral based Proto-Indo-Iranian religion (en.wikipedia.org/wiki/Proto-Indo-Iranian%20religion). According to Boyce (1975, 183) Zarathustra was trained as a priest in this religion, so like the other world religions we surveyed, the roots of Zoroastrianism lie in the oral tradition.

The Taoist religion grew out of the practices of the shamanistic tradition and only later did their sacred texts *Tao Te Jing* and *Zhuangzi* emerge. A similar pattern arose with Confucianism and the sacred texts of this religion. Confucius was a philosopher who drew his inspiration from the values and teachings of the Zhou dynasty that existed several centuries before his time. The four great Confucian texts of Great Learning, Analects, Mencius and Doctrine of Mean were not actually composed by Confucius. Great Learning was written by Zengzi, a disciple of Confucius. Analects was a compilation of Confucian sayings compiled by his followers. Mencius was written by the philosopher Mencius and Doctrine of Mean was written by his grandson Zisi.

Unlike the cyclic pattern of the oral tradition, the profane life of modern literate man follows a linear pattern of time like the lines of text that inform our lives. The sacred or religious life of modern humankind is based on the cyclic structure of time as it was during the oral tradition. The one exception might be the non-believers or secularist who adopt a scientific explanation for the origin of the universe and life. But even here it is possible that the sacred-like spirit invades their thought patterns. Science, we suggest, is another belief system trying to understand the sacred moments such as the beginning of the universe, the emergence of life, the emergence of human intelligence, and/or the origin of language. Science in the form of general systems theory, emergent dynamics and complexity theory no longer adheres strictly to a linear chain of cause and effect relationships based on reductionism. Rather a systems approach in which top down causality from a system to its components and a bottom up causality from the components of a system to the system as a whole operate simultaneously in a cyclic form of causality. It is like traditional religions in the sense that it is non-linear.

Chapter 10
The Spiral Structure of Marshall McLuhan's Thinking

In Australian Aboriginal and Maori literature, the circle and the spiral are the symbolic metaphors for a never-ending journey of discovery and rediscovery.
Eva Knudson (2004)
Cognition is re-cognition
McLuhan
Time present and time pastAre both perhaps present in time futureAnd time future contained in time past

Burnt Norton—T.S. Eliot

Introduction

We will make the bold assertion in this chapter that McLuhan's philosophy of media ecology has at its roots the spiral structure. We will further suggest that many of the artists, scholars and philosophers that had a significant influence on his thinking made use of spiral structures in their thinking as well.

We believe that a deeper understanding of McLuhan's life work and philosophy emerges by looking at the role of spiral structures in his understanding of media and culture. We are not suggesting that the spiral was foremost in his thinking but we believe that the archetypal structure of the spiral provides a frame in which a new view of McLuhan's work emerges and one that encompasses his reversals of figure and ground and that of cause and effect as well as the retrieval or flip in his Laws of Media (LOM). McLuhan always claimed that he had no theory but merely observed what people do and what were the effects of the tools, artifacts and technologies that mediated their interaction with the physical world, their social, political and economic world and their psychic and cultural world.

> I have no theories whatever about anything. I make observations by way of discovering contours, lines of force and pressures. I satirize at all times and my hyperboles are as nothing compared to the events to which they refer (Letter to William Kuhns December 6, 1971—Molinaro et al. 1987, 448).

R. K. Logan, I. Pruska-Oldenhof, *A Topology of Mind*, Mathematics in Mind,
https://doi.org/10.1007/978-3-030-96436-8_10

What is also interesting about his letter to Bill Kuhns is that McLuhan reinforces his claim that he does not start with a theory and that he admits explicitly that he makes use of satire.

The spiral structure of McLuhan's philosophical thought, we suggest, served as a stand-in for his lack of a formal theory. This hypothesis, in the spirit of McLuhan's work, is a probe and it may even be true or at least half true, which according to him was a lot of truth. He joked on a number of occasions: "There is a lot of truth in a half-truth (www.mediaeducationcentre.eu/eng/?page_id=442)."

We will first examine the spiral structure of the thinking and work of those scholars, philosophers and artists that most influenced McLuhan. We will then examine the spiral structure of the thinking and work of Marshall McLuhan himself, which we believe will provide a new way of viewing McLuhan's work. In particular, we will review the way he reversed figure and ground, content and medium, and cause and effect; the relationship he established between the content of a new medium and the older media it obsolesced; and the way time is structured of his Laws of Media (McLuhan, Marshall 1975, 1977; McLuhan, Marshall and Eric 1988).

Sources of McLuhan's Spiral Thinking

McLuhan's intellectual roots can be found in a number of other scholars, philosophers, literary figures and artists who embraced a spiral structure in their thinking and artistic productions including Giambattista Vico, Johann Fitche, Georg Hegel, Karl Marx, James Joyce, T.S. Eliot, Edgar Alan Poe, members of the Vorticism movement including Wyndham Lewis and Ezra Pound, Sigmund Freud, and I.A. Richards, McLuhan's professor in Cambridge where he did his PhD studies.

But what exactly do we mean by the spiral structure of a philosopher's thoughts or an artist's artistic output? The spiral structure of physical objects and biological phenomena is rather easy to identify. The spiral form of physical objects and biological structures represent a going back and forth in one dimension or one plane yet advancing in another dimension or direction. The application of the spiral form to thought processes, scholarship, and the structure of artistic and literary productions is metaphorical but yet very descriptive. The use of this metaphor provides two insights. It ties together the work of Marshall McLuhan with those that influenced his thinking and it provides another lens with which to view his scholarly output and his unique understanding of media and their effects. The spiral is a key that opens a door to a new perspective of McLuhan's work and his philosophy of media ecology.

The spiral structure of purely physical and biological phenomena is primarily played out in physical space, whereas the spiral structures of philosophy, culture, human thought, scholarship and artistic expression involve the time dimension. The movement back and forth in these domains entails the transitions from the present back to the past or forward into the future and vice-versa from the past and the future to the present. The spiral structure unites the past, the present, and the future. According to McLuhan, "We live in post-history in the sense that all pasts that ever were, are now present to our consciousness and that all the futures that will be are here now." He also suggested that, "the future of the future is the present (McLuhan 1999, 88)." Let us now examine the past with which McLuhan was familiar by exploring his acquaintance with the spiral structures of the philosophers, scholars and artists who directly influenced his spiral thought patterns.

Marshall McLuhan's Intellectual Roots

The foremost influence on McLuhan's interest with spiral structure was Giambattista Vico and James Joyce. It was probably Joyce that introduced the literary critic McLuhan to Vico's spiral notion of *corso* and *ricorso*. Giambattista Vico (1668–1744), an Italian grammarian, historian, political philosopher, jurist and author of The New Science (Scienza Nuova) developed a cyclic philosophy of history based on a society's period of advancement, corso, followed by a regression, ricorso, back to a more primitive stage. McLuhan pays homage to Vico and acknowledges his influence by titling the project that summed up his life's work: *Laws of Media: The New Science* (McLuhan 1988).

The influence of Vico on McLuhan and the spiral form of McLuhan's scholarship has been described by Sean Cubitt (2005) who wrote:

> In the work of the best-known among media historians, Marshall McLuhan (1911–1980), the narrative of evolution has an interestingly spiral dimension, returning in heightened form to its oral origin in the figure of the electronic "global village"... A noted Joyce scholar, McLuhan may have derived the spiral form of history from Joyce's source, Giambattista Vico (1688–1744). Be that as it may, he shares with other teleological media historians a mystical belief in eternal return or in millenarian thought that responds to a heartfelt longing for historical symmetry. McLuhan's spiral might be seen as syncretic, even atavistic.

James Joyce structures *Finnegan's Wake* using Vico's three ages into his narrative. As McLuhan observed, "The whole of *Finnegan's Wake* is a *ricorso*, a scrubbing purgation of private and corporate experience in the 'dreaming back' (McLuhan and Watson 2011, 148)." Joyce adds a fourth age to the cycle in the Wake, the age of chaos, which comes immediately after the democratic age of humans and just before the *ricorso*. Another parallel of Vico in Joyce is that both make use of etymology and rhetoric, and the significance they attach to myth. These parallels are also characteristic of McLuhan's scholarship.

McLuhan came to Vico through his analysis of James Joyce and in particular Finnegan's Wake as this excerpt from *From Cliché to Archetype*.

> Vico, in his *Scienza Nuova*, which Joyce found so useful, stresses that all ancient fables and tales are really records of moments of technical breakthrough to which the ancients assigned the status and name of a god, but Vico also insisted that the effects of such breakthroughs are recorded in new "wrunes" (line 36), writing into patterns of human speech and sensibility. Vico, like Joyce, insists that new technology is not added to culture, but it "ruins" whole societies, tossing them onto the midden hide or heap, whence they are forever being retrieved and refurbished by succeeding generations (McLuhan and Watson 2011, 102).

The influence of Vico directly on McLuhan has been described by Pietropaolo (1995):

> McLuhan and Vico share deep roots in the humanist tradition, from which they derive the delimitation of their chief objects of research, and both adhere without fail to a comparative approach aimed at the discovery of shared form in seemingly unrelated or distant phenomena. Thus, they identify large patterns of civilization and boldly illuminate them by overstepping the conceptual boundaries of individual disciplines and propounding instead all-inclusive science of men and communities. What is more, Vico was himself part of the tradition on which McLuhan relied in fashioning his own philosophical orientation and was therefore one of his theoretical sources. The paradigmatic presence of Vico is at times so conspicuous in McLuhan that a full exploration of his Vichianism, including that part of it, which McLuhan did not make explicit, would be a monograph requiring no justification whatsoever.

Another influence on McLuhan with a Vichian-like notion of *ricorso* was Lewis Mumford, a critic of architecture, industrialization and urban centers who believed the modern city did not serve the needs of its inhabitants. He advocated a return or *ricorso* to the ideal of the medieval city.

Like Vico, McLuhan believed that history could be used as a laboratory to understand the nature of human society. He was influenced by another historian, an economic one in the person of Harold Innis. In the Introduction to the second edition of the *Empire and Communication* McLuhan (1972a) wrote:

> [Innis] had discovered a means of using historical situations as a lab in which to test the character of technology in the shaping of cultures. Innis taught us how to use the bias of culture and communication as an instrument of research. By directing attention to the bias or distorting power of the dominant imagery and technology of any culture, he showed us how to understand cultures (McLuhan 1972a, xi).

McLuhan felt that an understanding of history was essential for understanding the future and the impact of new technologies. He often used the metaphor of the rear-view mirror; a device by which we are able to determine what is about to overtake us from our past. Furthermore, according to McLuhan, history is not to be regarded as a series of events but rather as a dynamic process with a discernible pattern, which repeats itself from culture to culture and from technology to technology.

A number of authors have suggested that McLuhan employed a dialectical approach to understanding historical processes. Grosswiler (1996) in an article entitled The Dialectical Methods of Marshall McLuhan, Marxism, and Critical Theory argued that, "McLuhan employed a form of dialectical theory containing basic elements of dialectics developed by Hegel, Marx, and, later, his contemporaries at the Frankfurt School." In fact, the dialectic method of thesis, antithesis and synthesis dates back to the work Johann Fitche. The influence of Fitche's triad of thesis, antithesis and synthesis can be felt in McLuhan's (McLuhan 1975, 1977, 1988) Laws of Media in which an artifact enhances some human function, obsolesces a former way of doing it, retrieve something from the past and pushed far enough flips into its opposite or complementary form. The enhancement element parallels thesis, obsolescence parallels antithesis and synthesis incorporates both retrieval and the flip.

The Spiral Structure of the Tetrad or Laws of Media

McLuhan's (McLuhan 1975, 1977, 1988) Laws of Media (LOM) also known as the tetrad represent McLuhan's figure/ground thinking. The LOM is sometimes formulated in terms of four questions and sometimes as four statements. We present both, first as four questions and then as four statements.

Four questions:

(a) What does a medium enhance?
(b) What does a medium obsolesce?
(c) What does a medium retrieve that had been obsolesced earlier?
(d) What does a medium flip into when pushed to the limits of its potential?

Four statements:

1. Every medium or technology enhances some human function.
2. In doing so, it obsolesces some former medium or technology, which was used to achieve the same function earlier.
3. In achieving its function, the new medium or technology retrieves some older form from the past.
4. When pushed far enough, the new medium or technology reverses or flips into a complementary or opposite form.

The enhancement as figure is to the obsolescence as ground as the retrieval as figure is to the ground of reversal.

The LOM represent a model for the evolution of artifacts. According to the LOM, every artifact when pushed far enough flips into a new more advanced artifact. As an evolutionary model, it explains the continuous emergence of new artifacts in the ongoing cycle of the four laws of enhancement, obsolescence, retrieval and flip. Each cycle of these four laws (the tetrad) is linked to the previous one and to the next cycle or tetrad and hence has a spiral structure. This linkage is illustrated for the evolution of communication media showing the transition from speech to pictographic writing; to the alphabet; to print; to computing; to the Internet; to the World Wide Web; to social media (Iseri and Logan 2016).

Artifact/ media Date Emerged	Speech 50,000 to 100,000 BC	Pictographic writing 3000 BC	Alphabet 1500 BC	Print block 600 BC Movable 1450	Computers 1945	Internet 1969	WWW 1990	Social media 1995
Enhancement	Human interaction	Memory and codification	Codification	Dissemination of alphabetic communication	Production of information	Dissemination of information	Internet with graphical displays	Social networking
Obsolescence	Mimesis	Speech/ oral communication	Pictographic writing	Hand written-manuscript	Hard copy information	Stand-alone computers	Text only messaging on Internet	E-mail
Retrieval	Co-operation	History, experience, heroic deeds	Speech/ oral communication	Greek culture through renaissance	Alphabet	Community	Billboards, televisions and movies	Insularity of a small village
Flip	**Pictographic Writing**	**Alphabet**	**Print**	**Computing**	**Internet**	**World Wide Web**	**Social media**	**Cyber bullying and gossip mongering**

Artistic and Literary Roots of McLuhan's Spiral Thinking and His Laws of Media

In addition to the influence of philosophers and historians we must also look at some of the literary and artistic influences on McLuhan's spiral thinking and his LOM or tetrad as well. One of the key ideas of McLuhan is the notion that with electric information we live in all times at once. It is an idea that he found in the poetry of T.S. Eliot, particularly in the Burnt Norton quartet of The Four Quartets, which begins with these lines:

Time present and time past
Are both perhaps present in time future
And time future contained in time past.
If all time is eternally present
All time is unredeemable.
What might have been is an abstraction
Remaining a perpetual possibility
Only in a world of speculation.
What might have been and what has been
Point to one end, which is always present.
Footfalls echo in the memory
Down the passage which we did not take
Towards the door we never opened
Into the rose-garden.

Allow us to speculate. Did McLuhan get the idea of the tetrad, the four Laws of Media of enhancement, obsolescence, retrieval and reversal or flip from T.S. Eliot's lines of poetry above? Or perhaps from the title of the series of poems titled *Four Quartets*, of which Burnt Norton is the first, written between 1936 and 1942 and published as a collection in 1943? *Four Quartets* consists of four poems interconnected with each other through their common themes of human's relationship with time, the universe, and the sacred. After all, the etymologies of the words tetrad and quartet basically signify "four," "a group of four" or "a set of four" (from Greek *tetrad* and from Latin *quartus,* respectively). Whether it was the title of this poem, or the title of the series of poems, or simply the first three lines of the first stanza that begin this poem and series, is difficult to say. However, the temporal dimension of these three lines is reflected in *Four Quartets* title itself, in which the repetition of the word of "four" in quartets through its shared etymology conveys the temporal dimension of language itself, as a living medium transformed in human mouths and ears.

Here is how we connected the first three lines from Burnt Norton in *Four Quartet* with McLuhan's Laws of Media, the tetrad:

Time present (enhancement) *and time past* (obsolescence)
Are both perhaps present in time future (flip or reversal)
And time future contained in time past (retrieval)

Additional evidence for the influence of T.S. Eliot's Four Quartets on McLuhan's formulation of the tetrad comes from an essay McLuhan (1968) wrote about the poem entitled "Rhetorical Spirals in Four Quartets". The title tells it all. In the essay McLuhan analyzes the four poems in terms of the five canons of rhetoric: invention, arrangement, style, memory and delivery. He then goes on to examine the Quartets in terms of the four levels of Biblical exegesis, i.e. the literal, moral, allegorical, and eschatological meanings of scripture. He also linked the Four Quartets to the spiral of the seasons suggesting Burnt Norton, East Coker, Dry Salvages and Little Gidding were associated respectively with autumn, summer, spring and winter.

Another spiral image that influenced McLuhan came from the short story of Edgar Alan Poe entitled, A Descent into the Maelström. McLuhan (1951) uses the image of Poe's maelström to describe in his preface to *The Mechanical Bride: Folklore of Industrial Man* the way the modern recipients of advertising can protect themselves from its insidious effects.

> Poe's sailor saved himself by studying the action of the whirlpool and by cooperating with it. The present book likewise makes few attempts to attack the very considerable currents and pressures set up around us today by the mechanical agencies of the press, radio, movies, and advertising. It does attempt to set the reader at the center of the revolving picture created by these affairs where he may observe the action that is in progress and in which everybody is involved. From the analysis of that action, it is hoped, many individual strategies may suggest themselves.

McLuhan made use of the maelstrom image throughout his career. Here is an excerpt from *Man and Media*, the last taped talk he gave in 1979 shortly before the stroke that ended his career.

> Our natural responses to media and to technology are irrelevant... we cannot trust our instincts or our natural physical responses to new things. They will destroy us. How are we to bypass or offset merely physical response to new technology and new environments created by new technologies?

> This problem has not been raised by anybody, even though we have to live with it every day. Edgar Allan Poe's story "The Descent into the Maelström" had tremendous influence on nineteenth-century poets and symbolists such as Baudelaire, Flaubert, and others. In this story, Poe imagines the situation in which a sailor, who has gone out on a fishing expedition, finds himself caught in a huge maelstrom or whirlpool. He sees that his boat will be sucked down into this thing. He begins to study the action of the maelstrom and observes that some things disappear and some things reappear. By studying those things that reappear and attaching himself to one of them, he saves himself. Pattern recognition in the midst of a huge, overwhelming, destructive force is the way out of the maelstrom. The huge vortices of energy created by our media present us with similar possibilities of evasion of the consequences of destruction. By studying the pattern of the effects of this huge vortex of energy in which we are involved, it may be possible to program a strategy of evasion and survival. (https://www.marshallmcluhanspeaks.com/lecture/1977-man-and-media/index.html, accessed May 7, 2021)

We suggest that McLuhan's writings are ideas that have reappeared and that by attaching ourselves to them we will find our way out of the digital maelstrom we are presently in.

Vorticism

Another influence of the spiral structure on McLuhan was his association with Wyndham Lewis and Ezra Pound, two members of the Vorticism movement.

> Vorticism might be described as visual energy… It involves creating layers within works, linking time and space, and basically sending the viewer or reader on a journey into a vortex. The best way to describe the "vortex" in Vorticism is by using the analogy of bifurcation points. When one places a pot of water and applies heat, the visible vortex is the point at which hot and cold clash to create change. When one pulls the plug from a bath tub, the point at which the surface tension breaks results in a vortex. It's the moment of change. Just like water in a draining tub, eventually the change runs its course and ends, which Vorticism did. Vorticism was a revolutionary artistic movement created to promote the *avant-garde* style. Some of its most noteworthy participants include: Wyndham Lewis, Ezra Pound and Henri Gaudier-Brzeska. (https://web.archive.org/web/20180727205854/https://peaceandwarpoetics.wikispaces.com/vorticism, accessed May 9, 2021).

McLuhan spent a great deal of time with both Wyndham Lewis and Ezra Pound, visiting with them and corresponding with them. In the collection of *The Letters of Marshall McLuhan* (McLuhan 1987) there are 40 letters to Wyndham Lewis between July 24, 1943 and August 12, 1956 and 22 letters to Ezra Pound between May 31, 1948 and January 22, 1957. In a letter to Lewis, McLuhan (1987, 160) acknowledges his debt to Lewis, "You have been, for years before I met you, a major resource in my life".

McLuhan also made extensive use of the following quotes from these two Vorticists: from Wyndham Lewis: "The artist is always engaged in writing a detailed history of the future because he is the only person aware of the nature of the present," which appears in *Understanding Media* (McLuhan 1964, 70) and from Ezra Pound: "The artists of our culture, 'the antennae of the race,' had tuned in to the new ground and begun exploring discontinuity and simultaneity," which appears in *Laws of Media: The New Science* (McLuhan 1988, 47). Both Lewis and Pound had a reputation for being notoriously anti-social. We believe that McLuhan had these two in mind when he wrote the following, "The poet, the artist, the sleuth—whoever sharpens our perception tends to antisocial; rarely 'well adjusted,' he cannot go along with currents and trends. A strange bond often exists among anti-social types in their power to see environments as they really are (McLuhan and Fiore 1967, 88)." And we suggest that a strange bond existed among the three anti-social types of Lewis, Pound and McLuhan. Although McLuhan was greatly influenced by Lewis and Pound he steered clear of their political views and focussed on their contributions as writers. It is also worth noting that McLuhan was also a fan of Lewis's paintings and actively sought commissions for his friend.

Elena Lamberti (2012, 148) in her book *Marshall McLuhan's Mosaic* points out the enormous influence of Wyndham Lewis on McLuhan's thought, "His [Lewis] works are among the first to conceptualize the idea of the global village, as well as the idea of tactility, two concepts that later became tenets of McLuhan's theories of media." McLuhan's admiration of Lewis is reflected in his pamphlet *Counterblast* (2011) originally published in 1954 that was "conceived in Lewis's vorticist style (Lamberti 2012, 215)."

A measure of the influence of vorticism on McLuhan was his many uses of the term vortex or vortices. He described media as "living **vortices** of power creating hidden environments (and effects) that act abrasively and destructively on older forms of culture (McLuhan 1972a, v, bolding is ours)." He made use of the metaphor of vortex to formulate his notion of the spiral structure in the reversal of cause and effect:

> The study of effects has lately driven me to the study of causality, where I have been forced to observe that most of the effects of any innovation occur before the actual innovation itself. In a word, a **vortex** of effects tends, in time, to become the innovation. It is because human affairs have been pushed into pure process by electronic technology that effects can precede causes. (Letter to Muriel Bradbrook November 5, 1971 National Archives of Canada McLuhan Collection, bolding is ours)

McLuhan also connects vortices and spirals: "The hidden borders in men's minds are the great **vortices** of energy and power that can erupt and **spiral** anywhere (McLuhan 1972b, bolding is ours)." And of course, McLuhan also invoked the image of the maelstrom of Edgar Allen Poe as we mentioned in the previous section.

Understanding McLuhan in Terms of Spiral Structures

We will suggest in the remainder of this chapter that a deeper understanding of Marshall McLuhan contribution to media studies and his Laws of Media (LOM) emerges by considering the spiral structure of McLuhan's thinking and specifically his LOM. This connection of the LOM and spirality was first suggested in print by Paul Levinson (1999) in his book *Digital McLuhan*,

> What radio obsolesces—visuality—television retrieves. And in so doing, television—what radio has flipped into—obsolesces the purely acoustic radio. There is a circularity of sorts here, which led me to term this ongoing movement of media "tetradic wheels of evolution." But actually… there is a real movement forward in this process—it is not just a circle—and so it might better be termed a spiral.

Levinson devoted an entire chapter of Digital McLuhan, Chap. 15—Spirals of Media Evolution, to the spiral nature of the Laws of Media (LOM). He actually first came up with the idea of the tetradic spiral at the Tetrad Conference he organized at Farleigh Dickinson University in March of 1978. That idea can be found in a typescript unpublished manuscript that can be accessed at https://www.academia.edu/28945044/Tetrad_Wheels_of_Cultural_Evolution (accessed October 17, 2017). Lindia and Levinson (2021) have returned to this topic is a paper entitled Tetrads and Chiasmus: A Reclamation of the Tetrad Wheel, which contains Levinson's previously unpublished manuscript just mentioned above plus an analysis of two articles that make use of Levinson's notion of the tetradic spiral, namely Curtis (1987) and Lindia (2020). The article further articulates the notion of the tetradic spiral.

Consideration of the spiral structure of McLuhan's thinking does more than just describe the LOM. It also throws light on the mystery of the McLuhan's success on two levels. Firstly, he did not work from a theoretical base yet he seems to have predicted or foreshadowed almost all of the technological developments of the digital age in this our Global Village. Something else allowed him to peer into the future, which we claim was the spiral structure of his thought patterns.

A second mystery requiring an explanation is his many self-contradictory remarks. We believe and hope to demonstrate that these remarks, which appear to be contradictory, actually represent the back and forth sides of the spiral structure of his thought processes in which by juxtaposing seemingly contradictory positions new insights emerge. We are not suggesting that the spiral was foremost in McLuhan's thinking but rather that his propensity to consider the reversals of (1) cause and effect, (2) figure and ground, (3) percept and concept, (4) cliché and archetype and (5) his use of the flip or reversal in his fourth law in his Laws of Media (LOM) reveals a pattern in his thought processes that resembles the universal archetypal structure of the spiral.

McLuhan Use of the Spiral Image in His Writings

We begin this probe by first collecting the many uses McLuhan made of the spiral image or structure in his writings. We have already cited his use of the image of the spiral form of Poe's maelström in the Preface to *The Mechanical Bride* and **xxx** the possible influence of the first three lines of Burnt Norton in his formulation of the tetrad or the Laws of Media.

In *The Global Village: Transformations in World Life and Media in the 21st Century* McLuhan and Powers (1989, 9) make explicit use of the spiral structure to describe the effects of the LOM or tetrad.

> The tetrad as a right-hemisphere visualization, helps us to see both figure and ground at a time when the latent effects of the mechanical age tend to obscure the ground subliminally. Its chief is that it raises the hidden ground to visibility enabling the analyst to perceive the double action of the visual (left hemisphere) and the acoustic (right hemisphere) in the life of the artifact or the idea. As such, the tetrad performs the function of myth in that it compresses past, present, and future into one through the power of simultaneity. The tetrad illumines the borderline between acoustic and visual space as an arena of spiraling repetition and replay, both of input and feedback, interlace and interface in the area of imploded circle of rebirth and metamorphosis.

We shall return to a more in-depth discussion of the tetrad or LOM, which demonstrates par excellent, the spiral structure of McLuhan's thought processes. But for now, we return to his explicit use of the notion of a spiral. In *Understanding Media* McLuhan (1964, 26) links the spiral structure of oral culture thought patterns with the similar concentric spiral pattern of the electric age as exemplified by Frank Lloyd Wright design of the Guggenheim Art Gallery:

> The Hebrew and Eastern mode of thought tackles problem and resolution, at the outset of a discussion, in a way typical of oral societies in general. The entire message is then traced and retraced, again and again, on the rounds of a concentric spiral with seeming redundancy. One can stop anywhere after the first few sentences and have the full message, if one is prepared to "dig" it. This kind of plan seems to have inspired Frank Lloyd Wright in designing the Guggenheim Art Gallery on a spiral, concentric basis. It is a redundant form inevitable to the electric age, in which the concentric pattern is imposed by the instant quality, and overlay in depth, of electric speed. But the concentric with its endless intersection of planes is necessary for insight. In fact, it is the technique of insight, and as such is necessary for media study, since no medium has its meaning or existence alone, but only in constant interplay with other media.

The next foray into McLuhan's explicit use of the notion of the spiral comes from his collaboration with the visual artist and filmmaker Sorel Etrog, who created an experimental film entitled *Spiral.* McLuhan and Etrog collaborated in the book *Excerpts from the Film Spiral by Sorel Etrog* with text and notes by Marshall McLuhan where McLuhan described the film in the following way: "In the film, *Spiral*, the ubiquitous and moving center intensifies awareness of the fragility and transience of existence…In *Spiral* duration is measured or fragmented. Chronological time yields to time as spaced-out moments of intensity… Everywhere in Spiral there is visually portrayed the labyrinth of the creative process (https://books.google.ca/books?id=rJl6IFuL1v0C&pg=PA110&lpg=PA110&dq=%22Spiral+duration+is+measured+or+fragmented%22&source=bl&ots=5zxNwLLS_Z&sig=ACfU3U0TAkeuRedTVBEFr_q0f8jc33Ny9A&hl=en&sa=X&ved=2ahUKEwj15Mu9z73wAhVCXM0KHbwnCIsQ6AEwAHoECAIQAw#v=onepage&q=the%20labyrinth%20of%20the%20creative%20process&f=false, pages 102–110, accessed May 9, 2021)." The project arose as a result of a screening of the film in one of McLuhan's classes after which McLuhan suggested that they create a book using stills from the film to which McLuhan would add appropriate text using lines of poetry from Eliot, Joyce and Yeats (Pruska-Oldenhof 2011).

The Implicit Use of Spiral Structures in the Thought Processes and Scholarship of Marshall McLuhan

In the last section we reviewed McLuhan's explicit use of the spiral in his writings. We now turn to the spiral structure of his thought patterns and his approach to his studies of media and their effects. Like the spiral form that proceeds in one direction and eventually reverses direction moving back and forth yet at the same time advancing in another dimension. McLuhan is constantly probing, trying out ideas, even ideas that seem to contradict each other. His thought patterns involve flips, reversals and contradictions yet somehow defying logic new insights emerge from this process.

Here is a set of his contradictory one-liners in which he takes his own ideas and probes with a grain of salt. These quotes can be found on the official website of the McLuhan Estate (https://www.marshallmcluhan.com, accessed May 9, 2021)

I may be wrong, but I'm never in doubt.
I don't necessarily agree with everything I say.
I've always been careful never to predict anything that had not already happened.
I don't pretend to understand my stuff. After all, my writing is very difficult
You don't like those ideas? I got others
Do you think my fallacy is all wrong?

Next, we present a group of one-liners (ibid.) in which he re-works a number of clichés providing new insights with his reversal of what is considered to be the figure and what is considered to be the ground. For instance, in the first example we present McLuhan reverses: 'necessity is the mother of invention,' which entails the idea that it is only after something is invented that it becomes a necessity.

*Invention is the mother of necessities (*https://exhibits.library.utoronto.ca/items/show/18171, *accessed May 9, 2021).*

The story of modern America begins with the discovery of the white man by the Indians.

[A shift from the ground of the European settlers to that of the indigenous people of America.]

Only puny secrets need protection. Big discoveries are protected by public incredulity.
I wouldn't have seen it if I hadn't believed it.
Mysticism is just tomorrow's science dreamed today.

Paradox, which played such a key role in McLuhan's work, is another form of a flip or the back and forth structure of the spiral.

Paradox is the technique for seizing the conflicting aspects of any problem.
Paradox coalesces or telescopes various facets of a complex process in a single instant.

In addition to these paradoxical and contradictory one-liners there are plenty of examples where McLuhan argues for what seems like contradictory positions in different places in his writings. He flips from one position to another spiraling back and forth all the while advancing and generating insights.

McLuhan on Artists

He held the artist in high esteem when he wrote:

> *I think of art, at its most significant, as a DEW line, a Distant Early Warning system that can always be relied on to tell the old culture what is beginning to happen to it.* (The *Distant Early Warning* Line [DEW] Card Deck [1969]—https://mcluhangalaxy.wordpress.com/2010/06/26/the-distant-early-warning-dew-card-deck-1969/)
>
> *The artist is the man in any field, scientific or humanistic, who grasps the implications of his actions and of new knowledge in his own time. He is the man of integral awareness* (McLuhan 1964, 71).
>
> *The serious artist is the only person able to encounter technology with impunity, just because he is an expert aware of the changes in sense perception* (McLuhan 1964, 33).

Like the structure of the spiral that goes back and forth but at the same time advances, McLuhan see the two sides of the contribution of the artist. The spiral structure appears in the feedforward and feedback of cybernetics and human culture. For McLuhan (1964, 70) the artist provides the feedforward, by "writing a detailed history of the future" The inventors and engineers, on the other hand, provide the feedback through the subliminal, counterintuitive effects of the technology they create.

Media as Extension of Man, Yet Man as Their Servomechanisms

Another juxtaposition is the contrast of McLuhan's notion that technologies or media are extensions of man but yet he saw its users as extensions of their technology as they become the servomechanisms of their own extensions. He expressed this idea as follows, "To behold, use or perceive any extension of ourselves in technological forms is necessarily to embrace it. By continuously embracing technologies, we relate ourselves to them as servo-mechanisms (McLuhan 1964, 46)." At first, technology serves as an extension of humankind serving our immediate needs but unbeknownst to us our tools slowly transform our environment and we become their servants or servomechanisms. Consider how the automobile has transformed our landscape especially in North America to suit the need of the automobile rather than the pedestrian. Once again, we encounter the spiral structure of McLuhan's thought as he asserts we become the extensions of our extensions.

The Figure and Ground Spiral

McLuhan's reversal of figure and ground is another spiral flip, which he expressed in the following excerpts from his writings,

> My writings baffle most people simply because I begin with ground and they begin with figure. I begin with *effects* and work round to the *causes*, whereas the conventional pattern is to start with a somewhat arbitrary selection of 'causes' and then try to match these with some of the effects. It is this haphazard matching process that leads to fragmentary superficiality. As for myself, I do not have a point of view, but simply work with the total situation as obvious *figures* against hidden *ground*. (Molinaro et al. 1987, 478).
>
> The figure is what appears and the ground is always subliminal. Changes occur in the ground before they occur in the figure. We can project both figure and ground as images of the future using the ground as subplot of subliminal patterns and pressures and effects which actually come before the more or less final figures to which we normally direct our interest (http://imfpu.blogspot.com/2008/12/magritte.html, *accessed May 9, 2021* and also available in the McLuhan file at the Canadian National Archive in Ottawa).
>
> In all patterns, when the ground changes, the figure too is altered by the new interface (McLuhan et al. 2003, 180).

A new medium or technology creates a new ground in which all the existing technologies operate and hence they are changed also.

> A new medium is never an addition to an old one, nor does it leave the old one in peace. It never ceases to oppress the older media until it finds new shapes and positions for them (McLuhan and Zingrone 1995, 278).

This the ultimate spiral structure as every medium or technology when pushed to its extremes flips into a new technology that is complementary or opposite to it as per the Laws of Media. But that original medium or technology is oppressed by the new medium it flipped into and hence the effects on it have spiraled back to it.

The Environment and Anti-environment Spiral

The reversal of figure and ground that we have just reviewed in the last section requires becoming aware of the ground, which is subliminal and hence is not always easy to identify. McLuhan suggests that in order to become aware of the ground or the environment one requires an anti-environment that only an artist or a scientist is capable of creating. In this process the environment that serves as the ground of the original figure of an artifact or a service flips into a figure operating in the ground of the anti-environment, thereby executing a spiral trajectory from ground to figure.

> Any new technology, any extension or amplification of human faculties given material embodiment, tends to create a new environment... It is in the interplay between the old and new environments that there is generated an innumerable series of problems and confusions... It is useful to view all the arts and sciences as acting in the role of anti-environments that enable us to perceive the environment (McLuhan and Zingrone 1995, 341).

In other words, it is the artist and/or the scientists by creating an anti-environment that allows us to perceive the new environment that is created by a new medium, which may be an artifact or a service. The anti-environment allows us to become aware of the environment that supports the new medium and not just the figure of that new medium as McLuhan (1970, 192) explains in Culture is Our Business:

> Since in any situation 10 percent of the events cause 90 percent, we ignore the 10 percent and are stunned by the 90 percent. Without an anti-environment, all environments are invisible. The role of the artist is to create anti-environments as a means of perception and adjustment. Hamlet's sleuth technique for coping with the hidden environment around him was that of the artist: "As I perchance hereafter shall think meet to put an antic disposition on" … (Act I, scene v)

One of McLuhan's favorite ways to describe our blindness to our environments was to point out that fish are unaware of the water they swim in. "One thing about which fish know exactly nothing is water, since they have no anti-environment which would enable them to perceive the element they live in (McLuhan et al. 1968, 175)."

The spiral structure is that of the content of a medium to the medium itself to the environment that the medium creates to the anti-environment that the artist or scientist creates that brings the subliminal environment/ground of the medium into our consciousness. McLuhan never used the image of a spiral to describe this process but we believe it is the appropriate structure to describe the twists and turns from content to medium to environment to anti-environment that the artist or scientist uses to bring the subliminal ground of the medium into view.

Cause and Effect Spiral

"Instead of asking which came first, the chicken or the egg, it suddenly seemed that a chicken was an egg's idea for getting more eggs" (McLuhan 1964, 12).

McLuhan (1964, 62) saw the creative process of both the inventor and the artist as working backwards from the effect they wanted to create to the cause that would lead to the desired effect.

> A. N. Whitehead, on the other hand, explained how the great discovery of the nineteenth century was the discovery of the technique of discovery. Namely, the technique of starting with the thing to be discovered and working back, step by step, as on an assembly line, to the point at which it is necessary to start in order to reach the desired object. In the arts this meant starting with the effect and then inventing a poem, painting, or building that would have just that effect and no other.

McLuhan explained how effects precede causes by showing how the effect of the telegraph was the cause of the telephone and the effect of the telegraph and the telephone was the cause of the phonograph.

> This revolutionary logic inherent in the electric age was made fairly clear in the early electric forms of telegraph and telephone that inspired the "talking machine." (McLuhan 1964, 279)

In the McLuhan approach, a medium or technology is studied by examining its effects not only on its immediate users but also on society as a whole. One of the ways of understanding the effects of a technology is to consider the changes that would take place if that technology suddenly disappeared. With McLuhan, the notion of a one-to-one correspondence between cause and effect gives way to a "total-field-theory" approach where one and the same cause may have a wide diversity of predictable and characteristic effects (McLuhan 1972, p. ix). Rather than "matching one cause against another effect [Innis] went on making sense of the process released by the new structures of communication (ibid., p. x)." Linear sequential cause and effect brought into doubt by Hume's skepticism and discounted as a principle of causality by quantum mechanics, is replaced by an interplay of causes and effects. McLuhan's work in this respect foreshadows complexity theory and emergence. The reversal of cause and effect is the technique of the artist who begins with the effect he or she wants to create in his or her audience and invents a work of art that achieves that end. The reversal of cause and effect is one of the consequences of electronically configured information patterns and the need for the cybernetic feedback and feedforward of information. The speedup of information flow requires planners to have complete knowledge of all possible ultimate effects, and hence the need for the reversal of cause and effect.

McLuhan deployed the reversal of cause and effect in his own analysis of understanding media and technology. He always started with the actual effects of media that he studied and not the mechanics of the technology that created them. He studied both the unintended and intended effects. The intended effects tend to be the services of the medium. The unintended effects are most often its disservices but as we have seen with the telegraph that is not always the case. One of the unintended effects of the telegraph was the telephone. And an unintended effect of the telegraph and the telephone was the phonograph and perhaps an unintended effect of the phonograph was the "talkies."

Service/Disservice

Another reversal of McLuhan is that of the service and disservice of a medium. Most analyses of media in the pre-McLuhan era focus on their service ignoring any disservices. Not McLuhan, he considered both as the excerpts from these two letters to colleagues indicate:

> All I am saying is that any product or innovation creates both service and disservice environments which reshape human attitudes. These service and disservice environments are always invisible until new environments have superseded them (To Jonathan Miller on April 22, 1970—Molinaro et al. 1987, 404).
>
> If you study symbolism you will discover that it is a technique of rip-off by which figures are deliberately deprived of their ground. You do not seem to have grasped that the message as it relates to the medium, is never the content, but the corporate effects of the medium as an environment of service and disservice (Letter to William Kuhns December 6, 1971—Molinaro et al. 1987, 448).

One can look at McLuhan's identification of the service and disservice of media as another example of his making use of figure/ground where the service, which was the intention of the medium, is the figure and the unintended disservice is part of the ground or environment that the medium creates. Once again McLuhan begins with the ground (the medium) and not the figure (the content) and focuses on the disservices of the medium unlike other communications scholars that pre-date McLuhan who almost always focused on the service of a medium ignoring its disservices. In the second excerpt he stresses the importance of the ground in understanding a figure and that the message of the medium is the ground it creates and not the content it delivers, which is why the medium is indeed the message. Once again, we have the reversal of the spiral structure. The service represents the progress of the medium and the disservice the regress. It has the same structure as Vico's *corso* and *ricorso*.

Two More McLuhan Figure/Ground Reversals

McLuhan's reversal of figure/ground resulted in two other spiral structures with two of his most famous one-liners. For example, in his notion that 'the medium is the message' it is the message that is the figure and the medium that is the ground. McLuhan spirals back to the medium that transmits the content as the more important message of a medium. The same figure/ground reversal is true of the expression 'the user is the content' where the content is the figure and the user is the ground. Each reader or viewer brings his or her own experience to a medium and transforms the content according to his or her own need.

Conclusion

We believe that the Laws of Media most vividly illustrate the spiral structure of McLuhan's thought processes that allowed him to pioneer the emergence of the new interdisciplinary, multidisciplinary field of the study of media ecology, a form of systems theory in which causality operates simultaneously top down and bottom up. The LOM represent the culmination of McLuhan's lifelong project to understand media and their impact on all aspects of human life. The spiral structure of the LOM mirrors the spiral structure of the internal workings of McLuhan's thought processes and is an important part of his legacy.

Spiral thought patterns **enhance** seeing both the liminal and the subliminal.

They **obsolesce** reductionist thinking.

They **retrieve** general systems theory, cybernetics and emergent dynamics.

And pushed far enough they will **flip** into the ultimate control of human's artifacts to serve human needs and the liberation of humankind's subservience to their technology as their servomechanisms.

Chapter 11
Spirals and the Arts

The artist is always engaged in writing a detailed history of the future because he is the only person aware of the nature of the present.
Wyndham Lewis (quoted on p. 65 of Understanding Media by Marshall McLuhan (1964))
Art at its most significant is a distant early warning system that can always be relied on to tell the old culture what is beginning to happen.
Marshall McLuhan (1969) Interview
Like Poe's mariner in 'Descent into the Maelstrom' (one can only) stave off disaster by understanding the action of the whirlpool. His insights offer a possible stratagem for understanding our predicament, our electrically configured whirl.

McLuhan and Fiore (1967, 150)

Introduction

This chapter is a contribution to previous attempts of scholars and artists to explore the spiral form as a universal structure and its manifestation in the arts. Texts that embark on exploring the spiral as an underlying form of nature, planetary movements, the structure of galaxies and nebulae, DNA, RNA, proteins, biological structures and its expression in human thought, language arts, fine arts, as well as science and philosophy, are often undertaken by people who dare to move beyond the specialist knowledge and one discipline to be able to apprehend the larger pattern at play. Over the last century, several books have been written on the spiral, all of which accorded significant attention to art and creative expression.

The first of such studies was by Theodore Cook (2009) in *Spirals in Nature and Art: A Study of Spiral Formations*, published in 1903. Cook was primarily known as a historian of specific regions of France (Touraine). However, he also studied biology and mathematics in Paris, which had opened to him other perspectives that

R. K. Logan, I. Pruska-Oldenhof, *A Topology of Mind*, Mathematics in Mind,
https://doi.org/10.1007/978-3-030-96436-8_11

specialization in one subject alone could never have achieved. Perhaps, this is what had attracted him to the manuscripts of Leonardo Da Vinci, who figures centrally in Cook's book. Cook's own work also moved in several directions, from science, biology, mathematics, and engineering to sculpture, painting, and architecture. And it was architecture (the open staircase at Blois, in Touraine, based on Da Vinci's designs) that brought Cook to look more closely at Da Vinci and to notice the form of the spiral, which he subsequently was able to generalize to nature and human.

Jill Purce (1971) first published a short journal article "The Spiral in Art: Aspirations Towards Eternity," which was followed in 1974 by her book *The Mystic Spiral: Journey of the Soul* (Purce 1980). Purce researched the form of the spiral and the theme of the labyrinth in nature, science, art, psychology and spiritual traditions. She identified a "linguistic connection between spiral, inspire, aspire, spiration, spirit and saw the spiral as an all-embracing structure (http://www.jillpurce.com/biography-of-jill-purce, accessed April 19, 2019)."

Nico Israel's book *Spirals: The Whirled Image in the Twentieth-Century Literature and Art* (2015) deals directly with manifestations of the spiral form in creative expression of twentieth century artists and authors. Israel's book explores this form in literature and several arts disciplines, it also broadens the scope of the study by including the art discipline that took over the twentieth century—cinema.

Poetry and Literature

Language reveals not only the contents of thoughts but also the patterns that reflect social dynamics and the dominant modes and media of communication. Oral or preliterate cultures rely on various mnemonic devices to remember and share stories that have been passed on from one generation to another. In his book *Orality and Literacy*, Walter Ong (1991) drew attention to repetition in the act of storytelling and the cyclical (spiral) structure of stories as common in oral cultures. He pointed out the presence of specific acoustic qualities, which empathized tonal and melodic qualities in language as well as rhythmic patterns that slowly disappeared with the introduction of the written word and literacy. The vestiges of oral cultures and the pre-literate past of our language still remain in poetry. Terms like lyric poetry and song reminds us of the strong presence of acoustic dimension in our language of the past, when poems were sung and accompanied with the sounds of lyres strummed by bards. Those tonal, melodic and rhythmic qualities of our past language remain in our studies of poetry (rhyme, metre, etc.) but as abstract and foreign notions to our present lived language. In the early twentieth century acoustic and sonic dimensions return to poetry and literature, not by following the established standards but by breaking them and breaking through into a new but familiar path, spiralling into a new environment.

Gertrude Stein: Repetition as a Spiral of Sensuous Revolutions

Repetition in everyday speech, puts us in touch or bears the marks of language of primary oral cultures, where repetition predominated. Our spoken and written language are quite different. When we speak, we sometimes repeat, it helps with mental recall in the moment. Gertrude Stein used these repetitions of spoken language as a model for her written descriptions and poetry. In fact, as Lynn C. Miller (1993, 154) notes, "Stein developed a method of written description from her observations of what people revealed about themselves in the repetitions of everyday speech." Her use of repetition, challenged the logic of written language and brought it closer to the oral, "most clearly apprehended in performance (ibid.)." Therefore, the sense of Stein's writing surfaces in the process of reading, ideally out loud, while listening with defocused attention and gathering sense evoked through sounds between each breath or pause, rather than that of particular words.

However, Stein maintained that there are no repetitions in her writing, rather what the reader might perceive as repetition is in fact "insistence." Insistence, not repetition, expresses existence in language. For Stein repetition appears as inanimate, as verbal descriptions of something devoid of existence, something passed and, in the past, when these things themselves are not existing (are not alive and present) in language. In her 1934 text, "Portraits and Repetition," she states: "There is only repetition when there are descriptions being given of these things not when the things themselves are actually existing....Then we have insistence, insistence that in its emphasis can never be repeating, because *insistence is always alive* and if it is alive it is never saying anything in the same way because emphasis can never be the same" (Stein 1990, 290–291, our emphasis). Stein's insistence has a spiral like structure in that a spiral also involves repetition but each time the spiral circles back it does so in a slightly different way. Stein's insistence is an example of this.

Stein's verbal insistence also bears close affinities with fragmentation of the visual plane and representation in the visual art of the time, in particular in the Cubist paintings of Picasso and Braque. She used language the way Picasso used his paintbrush. While not parting with representation, their works incited another sense through soliciting different engagements with them by their audience. Just as Stein's writing requires defocused attention to gather meaning from sonic qualities and connotation when it is read out loud, Picasso's paintings equally require defocused visual attention. In particular, his paintings *Ma Jolie* or *Accordionist* (both from 1911) from his late analytic cubist period, permit the figure and ground of the painting to meld together, as he intended, and allows one to grasp the entire visual composition in one glance. Experienced as a whole, we quickly realize that the meaning no longer resides only between clearly identifiable figures in the painting but the figures and ground as a totality, a whole environment. Humans, objects, and world are no longer separate or organized according to hierarchies, previously visually represented through figure-ground relations and other techniques to signify distinctions, but all are brought to the same, one plane; no matter how fragmented or multi-perspectival they might be. This coexistence, even non-differentiation, of humans, objects and world, are also the hallmark of Stein's writings.

It is, therefore, not surprising that both Picasso and Stein shared admiration for each other's work, each creating a portrait of the other: Picasso created a painting, *Portrait of Gertrude Stein* in 1905 and Stein created word portraits of Picasso, one in 1909 "Picasso," and the other in 1923 "If I Told Him: A Completed Portrait of Picasso". Stein's two portraits of Picasso demonstrate development of her distinct style as illustrated in this excerpt from the 1909 portrait of Picasso:

> This one always had something being coming out of this one. This one was working. This one always had been working. This one was always having something that was coming out of this one that was a solid thing, a charming thing, a lovely thing, a perplexing thing, a dis-concerting thing, a simple thing, a clear thing, a complicated thing, an interesting thing, a disturbing thing, a repellant thing, a very pretty thing. This one was one certainly being one having something coming out of him. This one was one whom some were following. This one was one who was working. (Stein 1990, 294)

Here are two more excerpts, these from the 1923 portrait of Picasso:

> If I told him would he like it. Would he like it if I told him?
> Would he like it would Napoleon would Napoleon would would he like it.
> If Napoleon if I told him if I told him if Napoleon. Would he like it if I
> told him if I told him if Napoleon. Would he like it if Napoleon if Napoleon
> if I told him. If I told him if Napoleon if Napoleon if I told him. If I told him
> would he like it would he like it if I told him.
> ...
>
> They cannot.
> A note.
> They cannot.
> A float.
> They cannot.
> They dote.
> They cannot.
> They as denote.
> Miracles play.
> Play fairly.
> Play fairly well.
> A well.
> As well.
> As or as presently.
> Let me recite what history teaches. History teaches.

(Stein, http://writing.upenn.edu/library/Stein-Gertrude_If-I-Told-Him_1923.html, accessed April 19, 2019)

In "Portraits and Repetition," Stein claims that remembering is not part of her process in writing her portraits, since her portraits aim to bring into existence or present the essence of their subjects (people, places, things) through insistence, which gives shapes to rhythm unique to each subject. Moreover, while propelling her argument of the difference between "repetition" and "insistence," she makes a direct connection between her writing and cinema. She states:

> Funnily enough the cinema has offered a solution of this thing [remembering]. By a continuously moving picture of any one there is no memory of any other thing and there is that thing existing, it is in a way if you like one portrait of anything not a number of them.... I was doing what the cinema was doing, I was making a continuous succession of the statement of what that person was until I had not many things but one... (Stein 1990, 293–294)

It is clear to see connections between the succession of frames on film to create movement and succession of sentences in Stein's portraits to create movement or animate (give life) to the subjects of her portraits. Stein turns to parallels between her writing and cinema, when arguing that in portraits her changing emphasis with each sentence in successive moments is not repetition, contrary to common perceptions.

> As I said it was if you like, it was like a cinema picture made up of succession and each moment having its own emphasis that is its own difference and so there was the moving and the existence of each moment as it was in me (Stein 1990, 307–308).

Cinema, through its succession of frames to create movement while being in movement, is about the present and being in presence of a thing existing on screen. Not memory but present, the existence of the thing moving. It is one portrait made up of multiple frames playing in succession, which are almost identical but not quite, there are tiny differences in each frame. This is why insistence rather than repetition is more apt in describing the succession of these frames in the sequence, where it becomes clear that the portrait is the whole sequence brought to life through film apparatus rather than each film frame being an individual portrait in itself. The construction of a film sequence, made up of numerous frames arranged in succession into a single composition, resembled Stein's technique of writing portraits: "I was doing what the cinema was doing, I was making a continuous succession of the statement of what of what that person was until I had not many things but one thing" (Stein 1990, 288). Lastly, both cinema and Stein's portraits rely on external energy sources to bring them to life: film composition on the film projector, powered by electricity; and portraits on the reader, reading it out loud. Therefore, like in cinema, where each frame is different (images vary from frame to frame, ever so slightly), her portraits are a succession of sentences with slight variations of words, not repetition but an ever-expanding spiral of rhythm, sounds, connotations and presence.

To conclude this section, we would like to turn to Stein's perhaps most modest yet best known little poem, consisting of a single line: "Rose is a rose is a rose is a rose." It first appears in 1913 in "Sacred Emily" (http://www.lettersofnote.com/p/sacred-emily-by-gertrude-stein.html, accessed April 25, 2019), and subsequently in six more of her texts, including in her 1939 book for children *The World Is Round* (Stein et al. 2013). This modest line captured the imagination of many artists, including filmmaker Stan Brakhage, you adapted Stein's views on writing to his filmmaking practice. He unravels the latent meanings, concealed by the visual artifice of written language, by bringing it to life through oral expression. In one of his regular radio programs, Test of Time, Brakhage (1982) discusses this poem:

> Everyone makes a joke of it and it's supposed simply to mean that a poem is a poem and a rose is a rose. But it also means, if you slur the language a little ... as you go round and

> round with that line, you get 'Rose,' the name of it, is 'arose,' a growth process, growing up, is 'eros,' 'e-r-o-s,' a symbol for sex, is 'arrows,' 'a-r-r-o-w-s,' which could be either cupids' arrows or death, is 'sorrows' also a symbol for death; and birth, sex and death in relationship to women is what that poem has always meant. So, it's not just a toy, delightful as it is as that (Brakhage 1982).

This one line illustrates the richness inherent in Stein's writing, its attention to sound and oral dimension of language, and its cyclical nature of slowly unravelling meanings in a dynamic spiral of sensuous revolutions through oral expression and its sonic resonances.

Cinema: "The Medium is the Message"

Spiral is inherent in the film medium itself. In some languages, the terms used to describe the activity of shooting a film with a camera, e.g., in French *tourner* and in Polish *kręcić*, which roughly translate into English as turning and winding respectively, are suggestive of the spiral shape. This notion is now lost with today's commercial cinema whose medium is not literally film but rather the digital medium.

Analog Cinema

Film

Duchamp's kinetic experiments with rotoreliefs and Anemic Cinema [optical illusions of rotoreliefs] return in OpArt movement in the '60s, which focused on paintings and eye creating movement, and also more recently in digital media and digital moving image such as Etrog's Spiral and Frampton's Zorns Lemma 1970 (https://www.youtube.com/watch?, accessed April 26, 2019).

Spiral enters film through visual representation, clearly demonstrated in Duchamp's *Anémic Cinéma* (1926), which consists of spinning concentric and eccentric circles, and through the form of the film, i.e., its overall temporal shape. Another example is Fernand Léger's 1924 film *Ballet Mécanique* (https://www.youtube.com/watch?v=yrfibt6Bkwc?, accessed April 26, 2019).

P. Adams Sitney (2002) in *Visionary Film: The American Avant-Garde 1943–2000* identifies the following films as having a spiral structure:

La Région Centrale by Michael Snow (https://film-makerscoop.com/catalogue/michael-snow-la-region-centrale and http://www.tiffcanadaonscreen.com/experimental-film-and-video/la-region-centrale?, accessed May 4, 2019). Sitney (2002, 359) describes it in Visionary Film: "the opening spiral of Snow's film, where the space which the viewer must study for the next three hours gradually discloses itself as the image feels its way from close, out-of-focus ground to horizon."

Meshes of the Afternoon (http://www.totalshortfilms.com/ver/pelicula/286?, accessed June 17, 2019) by Maya Deren. "It has an intricate spiral structure based on the repetition, with variations, of the initial sequence of the film, and it has a double ending. In the opening shot a long, thin hand reaches down from the top of the screen to leave a flower on a road. A young woman, played by Maya Deren, walks along the road, picks up the flower, and glimpses the back of a figure turning the bend ahead of her." (ibid., 7)

These films can be described as formalist films, where the form or the overall structure, architecture, and architectonics of the work takes precedence over their linear narrative. Sitney's (2002) description of formal cinema, is also a fitting description of films with a spiral form. They include the following:

Marcel Duchamp (1887–1968), *Anémic Cinéma* (various versions were made in 1920, 1923 and, finally, in 1926), made with the help of Man Ray and cinematographer Marc Allégret (https://www.youtube.com/watch?v=dXINTf8kXCc?, accessed May 9, 2019). It used ten discs with spirals (also called rotoreliefs because the discs bearing spirals when set in motion, for example on a record player, produced 3D visual effects or retinal illusions of convex and concave conical shapes) and nine discs with text in French in a spiral formation (linguistic puns and alliterations producing the effect of circular thinking). Ten 'Optical Disks' alternate with nine 'Disks Inscribed with Puns' Printed on the last frame of film: "Copyrighted by Rose Sélavy [handwritten], fingerprint and date: 1926".

(a) *Rotary Demisphere* (Precision Optics, 1925) related to the film (https://www.youtube.com/watch?v=8aI9_Nv5rz4; https://www.youtube.com/watch?v=6MpcOWSoFOc; https://www.youtube.com/watch?v=Nb3hXxWji-4; https://www.youtube.com/watch?v=1Z14XACOzEk; https://www.youtube.com/watch?v=vY_NQLJybWY; https://www.youtube.com/watch?v=IV_cjgCTmuM, accessed April 13, 2019)
(b) Connect with Pierre Henri Marie Schaeffer's electro-acoustic music using phonograph players, and the connection between spirals in phonographs and cinema (https://www.youtube.com/watch?v=a6oaCG4TmFI, accessed April 17, 2019)

Amy Greenfield's Dervish (1972; http://film-makerscoop.com/catalogue/amy-greenfield-dervish-2) and Wild Fire (https://vimeo.com/73679482, accessed April 23, 2019).

Sorel Etrog born in 1933 in Romania and deceased 2014 in Canada was primarily a sculptor but he created one film, the 16 mm 30-min film *Spiral* with music by Dmitri Shostakovich in 1974 (see https://www.facebook.com/watch/?v=10155363913609754 for a small clip of the movie). Jill Purce (1980, 56) wrote, "In almost all religions and mythologies, the spiral always comes between man, death and rebirth," which connects with the central themes of Etrog's film Spiral. Together with Marshall McLuhan Sorel Etrog co-authored the book *Images from the Film Spiral* (https://books.google.ca/books?id=7A_pITY2HKAC&printsec=frontcover&source=gbs_ge_summary_r&cad=0#v=onepage&q&f=false, accessed April 26, 2019).

Zorns Lemma by Holis Frampton, which can be viewed at (https://www.youtube.com/watch?v=ec2-GZRYGd8, accessed June 6, 2019), "is a 1970 American structural experimental film… Originally starting as a series of photographs, the non-narrative film is structured around a 24-letter Latin alphabet (https://en.wikipedia.org/wiki/Zorns_Lemma_(film), accessed June 6, 2019)."

Robert Smithson created the film Spiral Jetty (https://pl.khanacademy.org/humanities/ap-art-history/later-europe-and-americas/modernity-ap/v/smithson-jetty) that documents Smithson's land art of the same title. He wrote about his film, "For my film (a film is a spiral made up of frames) I would have myself filmed from a helicopter (from the Greek *helix, helicos* meaning spiral) directly overhead in order to get the scale in terms of erratic steps" (Smithson 1996, 143–153).

"Robert Smithson's earthwork *Spiral Jetty* (1970) is located at Rozel Point peninsula on the northeastern shore of Great Salt Lake. Using over six thousand tons of black basalt rocks and earth from the site, Smithson formed a coil 1500 feet long and 15 feet wide that winds counterclockwise off the shore into the water (http://www.diaart.org/sites/main/59, accessed June 9, 2019)."

Music and Sound Art

Karlheinz Stockhausen, *Spiral* (1968) or No. 27. Composition for a soloist and short-wave receiver [3 microphones, 2 × 2 loudspeakers, mixing console/sound projectionist, who also controls the lighting from the mixing console in the hall] (http://www.karlheinzstockhausen.org/complete_list_of_works_english.htm, accessed May 17, 2019). Its world premiere was in Zagreb, 1969. It was performed 1300 times at the 1970 World Expo in Osaka in 1970 (http://www.medienkunstnetz.de/works/stockhausen-im-kugelauditorium/images/4/, accessed May 17, 2019).

Pierre Henri Marie Schaeffer (August 14, 1910–August 19, 1995) was a French composer, writer, broadcaster, engineer, musicologist and acoustician. Electro-acoustic music and use of turn tables (see Figs. 11.1 and 11.2). From *In Search of a Concrete Music:*

> The paradox was that for two years I had been practicing concrete music but without having yet discovered it. I had discovered operational procedures, I was capable of manipulating, and I was nowhere near being as advanced on the theoretical level. I was a prisoner of my closed grooves…. Before becoming a method, it appeared as a "gimmick," a sound "effect." But from being an effect, it can become a cause and a means of discovery. The latter arises from a symbolic difference: the difference between a spiral and a circle. It turns out that the disc cutter is a machine that draws its own symbol. The cutter's spiral is not only the material realization but also the affirmation of time going by, time gone by, which will never come again.

Steve Reich is a minimalist composer strongly influence by a variety of composers ranging from J.S. Bach Debussy, Bartok and John Coltrane. His music contains repetitions and spiral structures. He is quoted as saying:

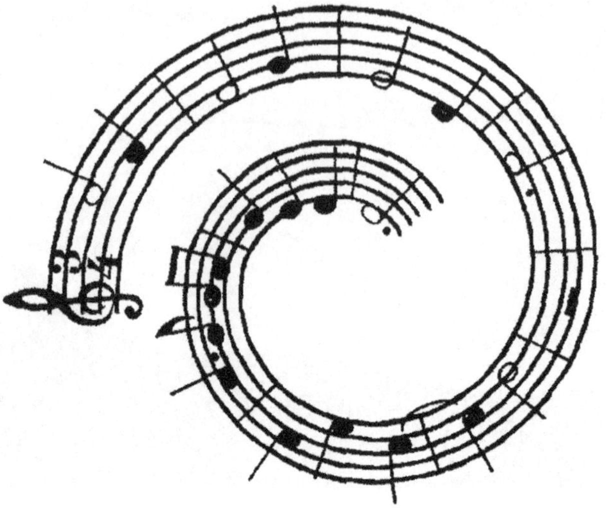

Fig. 11.1 Recording spiral (symbolic representation)

Fig. 11.2 Closed groove (symbolic representation)

> The cyclical is only interesting when it's not a cycle but when it's a spiral. If it goes around and around in a circle, you're really a rat in a trap, and just playing a loop is a bore. But if you return … above that point, or in a different position, you have returned as a different person, you have returned as a different composer, and you have returned to a different musical accomplishment (https://www.theguardian.com/music/2016/oct/26/steve-reich-interview-carnegie-hall-composer, accessed May 21, 2019).

Philip Glass is another minimalist composer in which makes use of spiral structures.

> Glass himself never liked the term and preferred to speak of himself as a composer of "music with repetitive structures." Much of his early work was based on the extended reiteration of brief, elegant melodic fragments that wove in and out of an aural tapestry. Or, to put it another way, it immersed a listener in a sort of sonic weather that twists, turns, surrounds, develops (https://web.archive.org/web/20130804233252/http://www.philipglass.com/bio.php, accessed May 21, 2019).

Alight Spiral Snip is an example of a Glass composition with a spiral structure (https://www.discogs.com/composition/189b518a-6010-4d70-9515-5d8244f0060a-Alight-Spiral-Snip, accessed May 21, 2019).

An Example of Spiral Structures in Dance: Loïe Fuller and Martha Graham

No discussion of the spiral form in the arts can fail to mention two dancers: Loïe Fuller (1868–1928) and Martha Graham (1894–1991).

Fuller created a unique form of dance in which she created spiral forms using her voluminous silk costume (see the accompanying photo) that completely concealed her body and that was illuminated with a special lighting of continually changed colors. The reader is advised to see the following YouTube video (https://www.youtube.com/watch?v=BU-k43TRlSE) which shows a dancer Papinta recreating Fuller's Serpentine Dance.

Martha Graham performed and a form of modern dance that completely broke with the rigidity of classical ballet. She created a technique of modern dance what came to be known as the Graham technique.

> The fundamental movement of Graham technique is the cycle between "contraction" and subsequent "release", which developed as a stylized representation of breathing… The second fundamental concept in Graham technique is the spiral The basic "spiraling" position consists of rotating the spine approximately 45° around its vertical axis, so that a dancer facing the front of the stage would have their shoulders aligned with the "Via Triumphalis", an imaginary line parallel to a corner-to-corner diagonal of the stage (https://en.wikipedia.org/wiki/Graham_technique, accessed January 25, 2020)

1. Graham Glossary

   The Graham contraction: Deep contraction originating at the pelvis that causes the torso to react in a curve.
   The Graham spiral: Twisting of the torso around the spine (https://www.dancespirit.com/martha-graham-how-to-2326135857.html, accessed January 25, 2020).

2. Martha Graham Technique Spiral (https://www.youtube.com/watch?v=E8p9Fpyv4ng, accessed January 25, 2020).
3. Definition of Martha Graham Technique

Three signature moves define Graham technique: the contraction, the release and the spiral. The Graham contraction is a C-curve of the spine. The release is the response to the contraction, which returns the spine to a neutral position or opens the breastbone to the sky. The spiral is a twisting of the torso that starts at the pelvis and then moves up the body (https://prezi.com/vwbzvn-hyjcrd/martha-graham-technique-journal/, accessed January 25, 2020).

Spiral Structures in Twentieth and Twenty-First Century Architecture

In the past 100 years many architects have incorporated spiral structures into their design of buildings and elements within their buildings. Here are some examples.

Vladimir Tatlin's *Monument to the Third International* (1919–1920)—a grand monumental building that were to be built in St. Petersburg after the Bolshevik Revolution in 1917. It was never built and remained as models and sketcher. The main form of this structure was a double helix that spiraled from wider at the bottom to narrow at the top.

Frank Gehry made use of spiral structures in three of his projects. The Art Gallery of Ontario's spiral staircase *Baroque Stair*, begins in the Walker's Court as an all-wood staircase and juts outside, covered with metal and glass, facing the Grange Park. This spiral staircase is irregular rather than logarithmic, i.e., self-similar spiral curve. It was completed in 2008. Matthew Teitelbaum noted that Gehry designed the Baroque Stair to be a "place of experience, not just a way of getting somewhere." (http://www.ago.net/baroque-stair, accessed March 4, 2020). He also designed a nine-story building, the Gehry Tower with a spiral shape to its form in Hannover Germany (1999–2001). A third use of a spiral structure was incorporated in the 12-story Opus apartment building in Hong Kong completed in 2012.

Baroque Staircase

Gehry Tower—Hannover

Opus Apartment—Hong Kong

Le Corbusier or Charles-Édouard Jeanneret-Gris. Le Corbusier's spiral plan projects include: sketches for the Villa La Roche (1923), Museum of Unlimited Growth (1939).

Villa La Roche

Museum of Unlimited Growth

Frank Lloyd Wright—The Solomon R. Guggenheim Museum, New York City (1943–1959).

> Continuous spiral ramp inside the building around the edges takes gallery visitors from the bottom to the top floor of this building, ending at the ceiling skylight. Cylindrical building with a spiral shape outside (the façade) and inside the building, the ramp. His inverted ziggurat (a stepped or winding pyramid of Mesopotamian origin) dispensed with the conventional approach to museum design, which led visitors through a series of interconnected rooms and forced them to retrace their steps when exiting…. The galleries were divided like the membranes in citrus fruit, with self-contained yet interdependent sections. The open rotunda afforded viewers the unique possibility of seeing several bays of work on different levels simultaneously. The spiral design recalled a nautilus shell, with continuous spaces flowing freely one into another. (Matthew Drutt, https://www.guggenheim.org/the-frank-lloyd-wright-building, accessed March 4, 2020)

View of spiral ramp

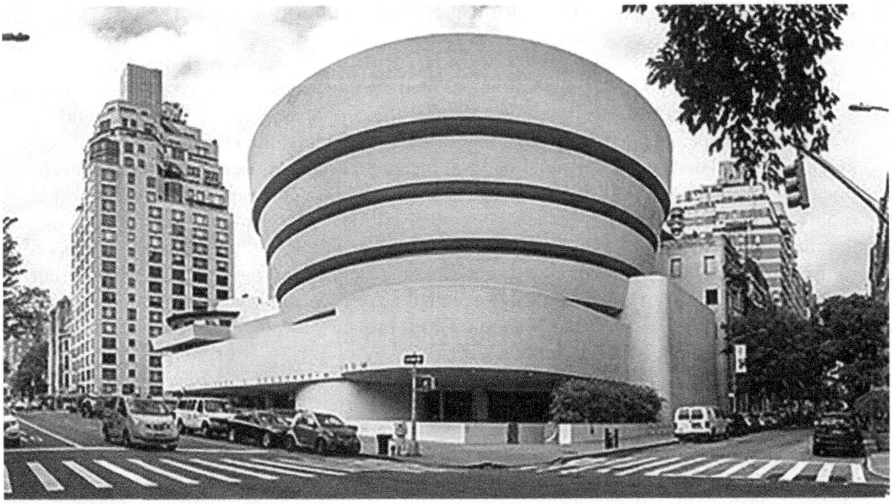

View from Fifth Avenue

Twist Scrapers

Santiago Calatrava, architect. Turning Torso in Malmo, Sweden. The tallest skyscraper in Sweden. Completed in 2005. A helical vertical shape, i.e., rectangular building that twists 90°, a 90° spiral (https://www.arch2o.com/turning-torso-santiago-calatrava/, accessed March 4, 2020, Courtesy of James Silverman).

Turning Torso

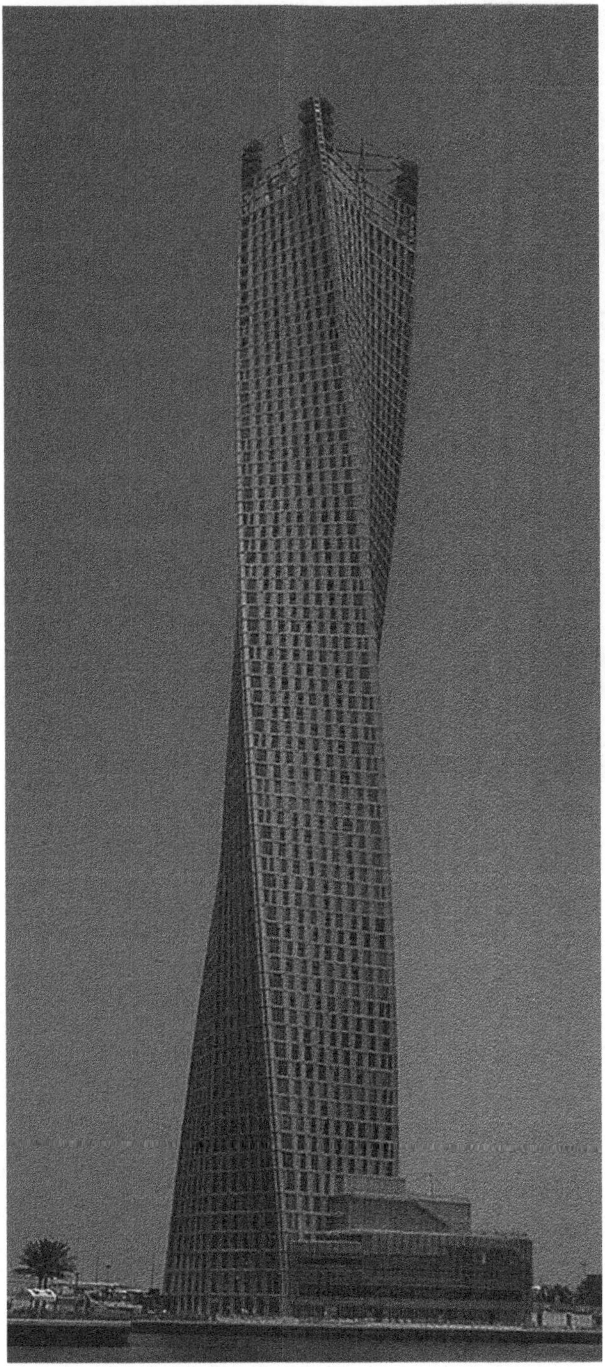

Cayan Tower

Skidmore, Owings and Merrill SOM architectural group. Cayan Tower (formerly Infinity Tower) in Dubai (began in 2006, completed in 2013).

Faruk Kapidžić Avaz, architect. Twist Tower in Sarajevo, Bosnia and Herzegovina (2006–2008). A twisted, helical façade.

Twist Tower

Absolute World Towers

Burka Architects and MAD Architects. *Absolute World Towers* in Mississauga, Ontario, Canada nicknamed Marilyn Monroe. Two towers twist 209° angle. Tower one began in 2007, tower two in 2010; 2007–2012.

Other towers with a twist to them include:

Kuwait Trade Centre (Al Tijaria Tower) in Kuwait City. Helical twist (2009) (http://www.skyscrapercenter.com/building/al-tijaria-tower/1373, accessed March 4, 2020).

Revolution Tower in Panama City. 52-floors that twist 360 degrees, like a corkscrew (2011) (https://en.wikipedia.org/wiki/F%26F_Tower, accessed March 4, 2020).

Generali Tower in Milan designed by Zaha Hadid Architects will be among several building that will comprise CityLife Milano masterplan (approx. a 45-degree twist) (https://en.wikipedia.org/wiki/Generali_Tower, accessed March 4, 2020).

Evolution Tower, Moscow (2011–2014), designed by Tony Kettle. A twist of 135 degrees (https://en.wikipedia.org/wiki/Evolution_Tower, accessed March 4, 2020).

Mode Gakuen Spiral Towers, Nagoya City in Japan. (2005–2008) Towers are wing-like and as the building rise they become narrower at the top, while also changing the rotation of the axis (https://en.wikipedia.org/wiki/Mode_Gakuen_Spiral_Towers).

Kuwait Trade Centre

Revolution Tower

Generali Tower

Evolution Tower

Mode Gakuen Spiral Tower

Spiral Staircase—Home Depot (https://www.homedepot.com/p/Arke-Civik-55-in-Black-Spiral-Staircase-Kit-K03018/100663970, accessed March 4, 2020).

Sculpture

Sorel Etrog whose film Spiral we discussed above was a sculptor and some of his pieces have a spiral-like structure to them:

Homage to Louis Armstrong

War Remembrance

Source 1

Toronto-based **Mark Birsted** creates hand-bent metal sculptures

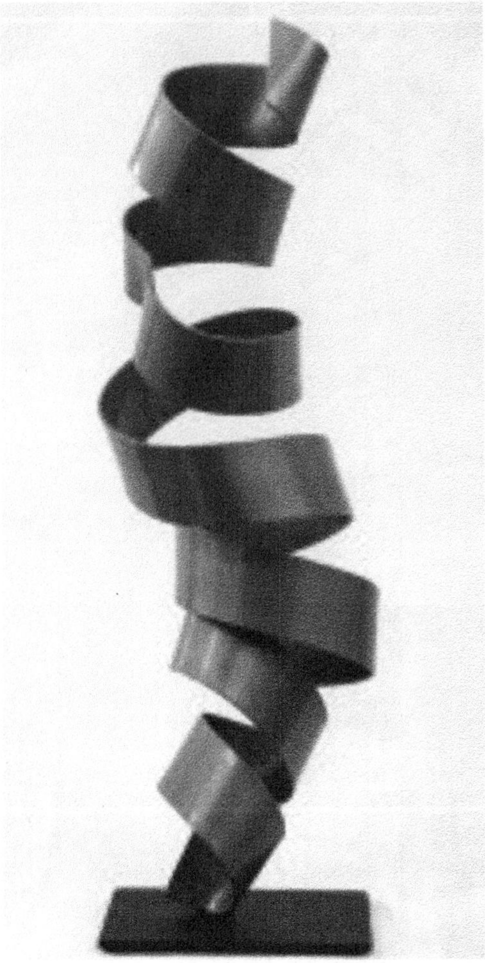

Hope Silver

Recline Blue

Richard Serra's choice of the material (weathering steel) further contributes to the temporal quality of his sculptures, which is experienced while traversing between the walls of its interior and when looking at the rust settling on the steel.

Forty Years

Inside Out

Torqued Spiral

Mike Smith 2005 Remake of Continuum (1963), a sculptural installation, acrylic on aluminum panels which is no longer extant.

Continuum (Remake) (2005) acrylic on aluminum panels, 209 × 275 × 361.8 cm. This is an actual spiral environment that requires people to enter and be immersed in it.

Continuum (Remake) - Mike Smith Studio

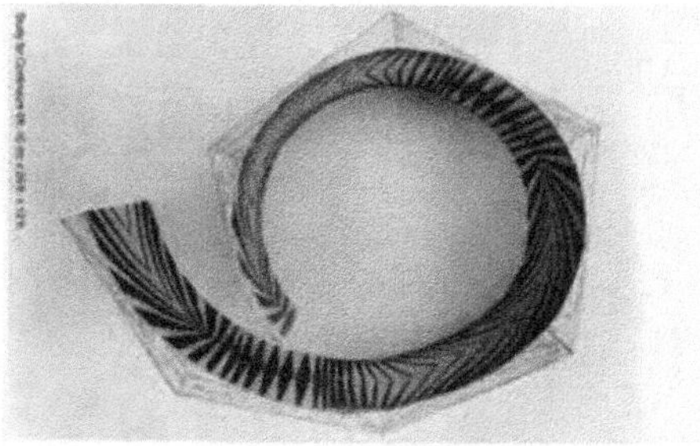

Continuum (Remake) - Mike Smith Studio

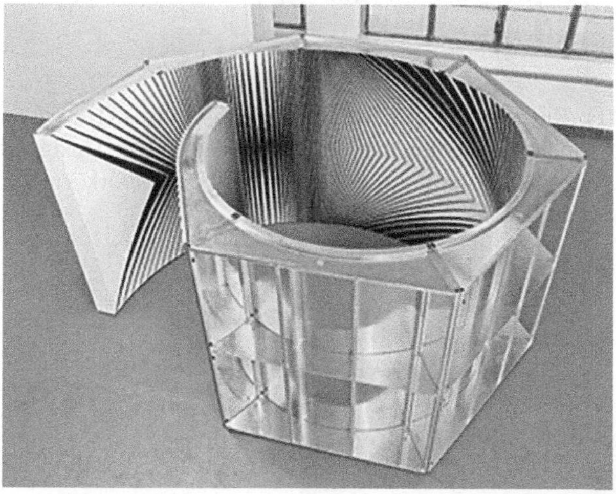

Visual Arts

Gustave Klimt, Lebensbaum (The Tree of Life) (1909), a three-panel mosaic frieze (1905–1911). Commissioned by Brussels-based Stoclet Palace.

El Lissitzky (1890–1941), Russian avant-garde (Suprematist) artist. He was a visual artist, designer, photographer, typographer, polemicist and architect. Designed art exhibition displays and propaganda works for Soviet Union. His work greatly influenced constructivist movements and Bauhaus.

Free-Floating Spiral (1923) graphite and colour pencil, India ink, pen and gouache on paper.

Francisco Infante-Arana (born 1943 in Vasilievka, Russia).

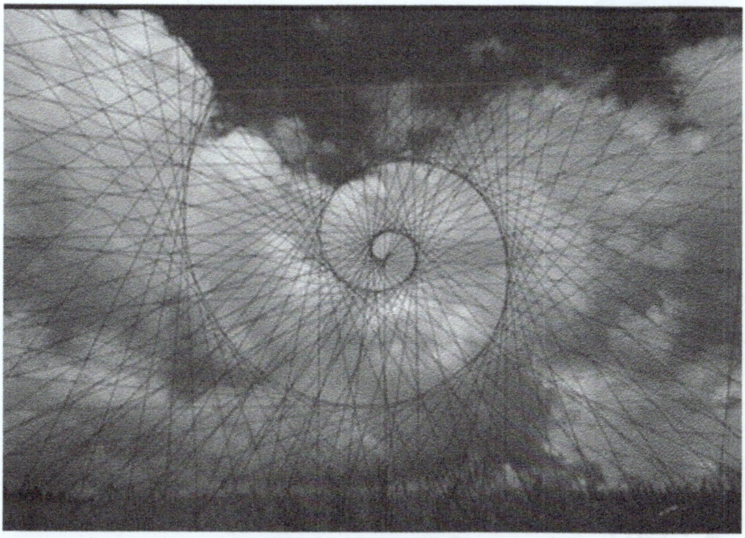

Lighter Spaces

Lighter Spaces

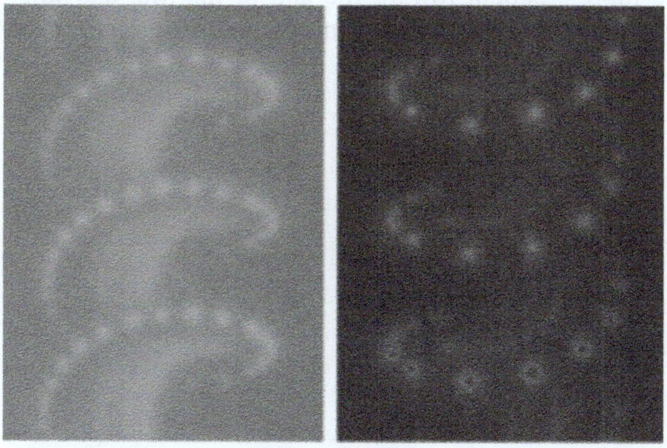

Spiral Lights

Spiral Lights

Hell of a Gap

Alexander Calder's (1898–1976) paintings from 1960s and 1970s include the spiral motif. These become more frequent toward the end of his life, and interspersed among geometric and biomorphic forms (http://www.artnet.com/artists/alexander-calder/2, accessed April 4, 2019).

Grande spirale
en couleur

Grande spirale en couleur4

Spiral with pumpkin

Spiral with pumpkin

Untitled

Untitled

Bridget Riley (born 1931 London, England).

Even though she did not paint spirals per se, several of her Op-art works, in particular those using circular and oval shapes, when looked at produce the illusion of moving spirals, not dissimilar from Duchamp's rotoreliefs when spun on the record player, which were after all made of variously placed circles descending in size toward the centre. Inspired by impressionist and post-impressionist painters at the turn of the twentieth century and their interest in new optical effects based to a great degree on the scientific discoveries in physiology of the eye and vision, as well as the new photochemical means of producing images through photography, Riley's interest in optical phenomena in art began quite early in her career as a painter. Here is an example of her work that incorporates a spiral structure, Blaze (1964), a screen print on paper:

"The defining exhibition in the history of Op Art," and one of the major contemporary art events of the decade, was MoMA's 'Responsive Eye' exhibition from 1965. The exhibition had been conceived (although not titled) and announced several years earlier in 1962, with a view to documenting the development from Impressionism to the exciting new trend of a "primarily visual emphasis (http://www.op-art.co.uk/Richard%20Anuszkiewicz/, accessed March 15, 2019)."

Other Op-art artists whose explored spiral and cyclical forms as dynamic in painting include:

Francis Celentano who wrote the following about his art:

> For me, so called Op Art or better Perceptual Art functions as a metaphor for the distortions of perception, experience and reason generously provided by nature and culture (http://franciscelentano.com/news.html, accessed February 21, 2019)
>
> These Kinetic paintings were in part inspired by my take on surrealist concepts that inform aspects my position in art up to the present. I designed, constructed, and timed each of their motors so that undulating movements would radiate from each surface creating a pulsing rhythm. The repetitive rotations seem almost organic and alive, generating a hallucinatory and destabilizing experience" Francis Celentano (http://franciscelentano.com/section/360467-Kinetic-Painting-1967-68.html, accessed February 21, 2019)
>
> I was curious about how I could transfer the tension I had gotten in color—whether contrast or harmony—to three dimensions while also working with the vertical format. I got the idea for a column that twisted in space. I was fascinated by achieving that spiraling movement (http://franciscelentano.com/artwork/3150908-Star-10.html, accessed February 21, 2019).

And here are some samples of Celentano's work from https://franciscelentano.com/section/360467-Kinetic-Painting-1967-68.html

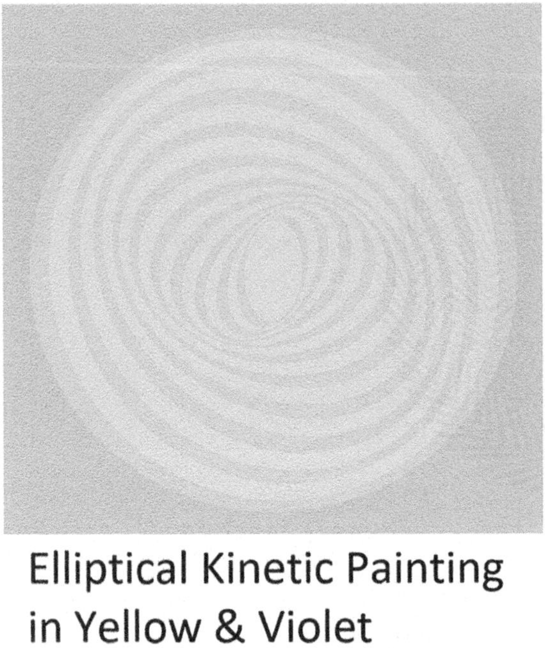

Elliptical Kinetic Painting in Yellow and Violet

Elliptical Kinetic Painting

Hexagonal Metamorphosis

Alice Aycock: *Project for a Circular Building with Narrow Ledges for Walking* (1976) reinforced concrete, cast in place, wood, overall height 17′, 13′ above grade, 4′ below, exterior diameter 12′; *Maze* (2000) aluminum 19.5″ × 31″ (http://www.lorettahoward.com/artists/aycock, accessed June 18, 2019).

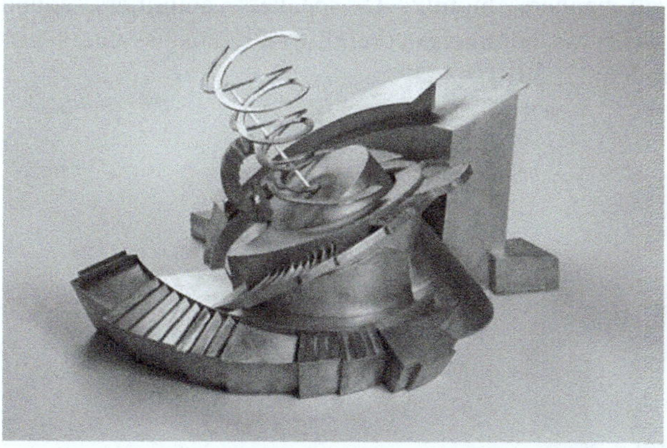

Judy Pfaff (born 1946 London, England), *Alberta* (2014) Steel, acrylic, resin, various plastics, florescent lights, expanded foam. Dimensions variable (https://issuu.com/lorettahoward/docs/judypfaff_issu/1?e=2803754/10006750, accessed April 13, 2019).

Sandra Meigs, *Room for Mystics* (2017) by Sandra Meigs and Christopher Butterfield30 paintings, wall banners, mobile, 15 speakers, amplifiers, sound (recorded and live) (http://www.sandrameigs.com/all-to-all/faickaisvmcntgsxjs-n62frlu0jlxv, accessed February, 27, 2019).

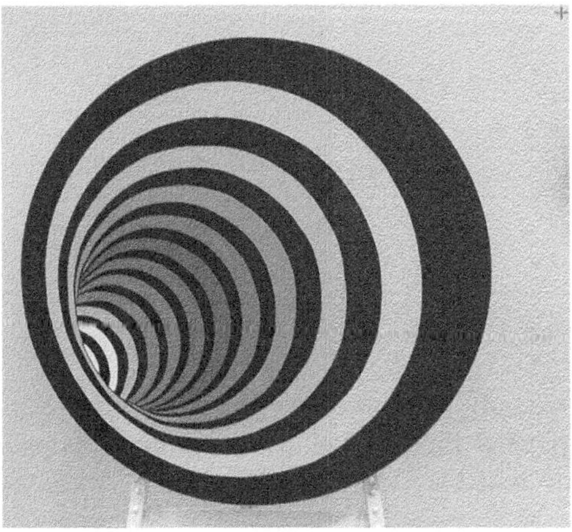

Mystic No. 4

Mystic No. 4

Mystic No. 3

Mystic No. 3

Photography

Man Ray made his "rayographs" without a camera by placing objects-such as the thumbtacks, coil of wire, and other circular forms used here-directly on a sheet of photosensitized paper and exposing it to light (https://www.metmuseum.org/art/collection/search/265487, accessed March 2, 2019).

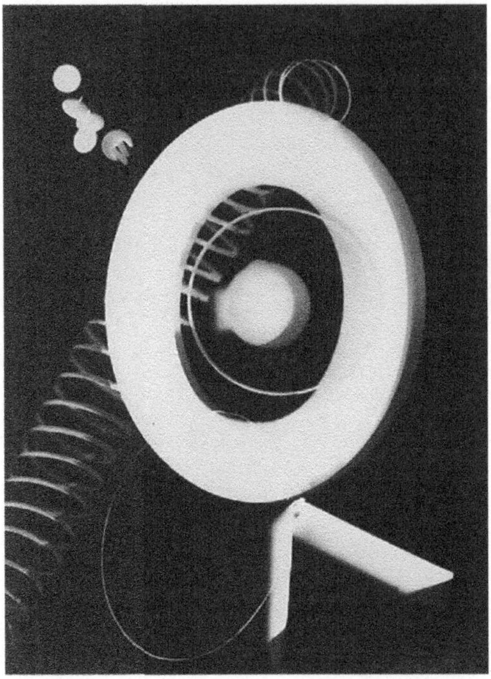

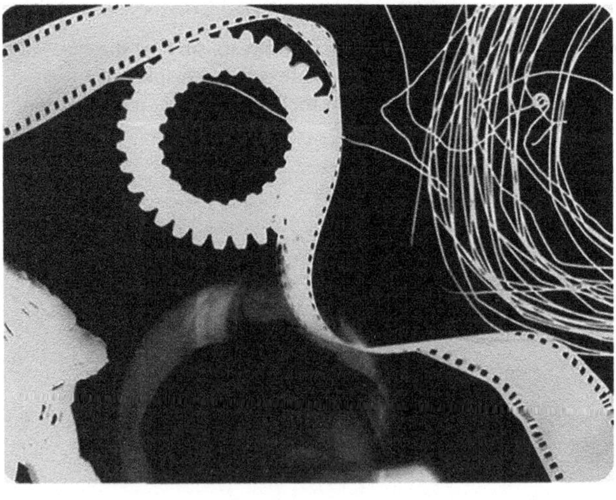

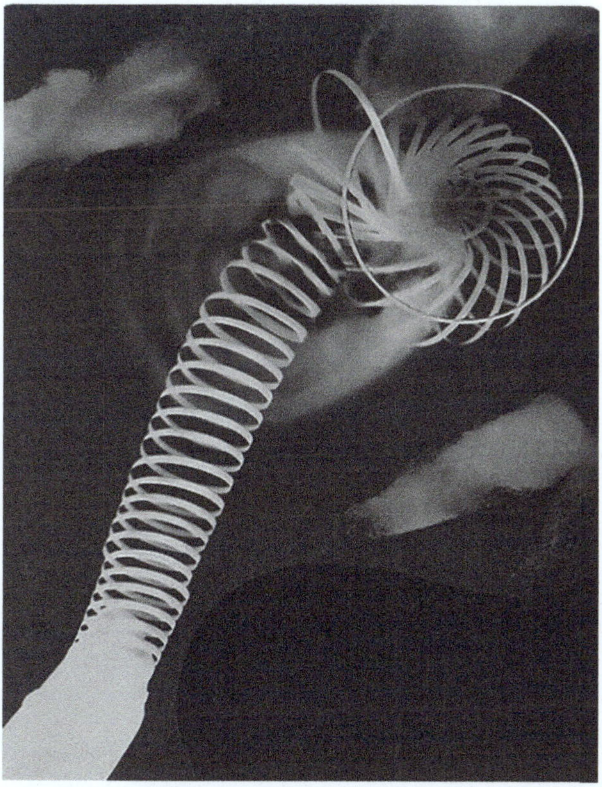

Prologue to Part III
Hyperlinking as Patterns of Connection

The universe is a hyperlinked connection of phenomena, things, interactions, and the like. All things are networked. In Part III of our study on the topology of the mind we will examine the nature of the links between different forms of information and phenomena both material and nonmaterial; profane and sacred; and textual (hypertexted) and non-textual (hyperlinked). We will do so by making use of McLuhan's figure/ground analysis derived from gestalt psychology in which the meaning of a figure depends on the ground in which it operates. Every figure has its ground and operates in some environment. And no environment stands by itself, but is connected to other environments. There is nothing in the universe that stands by itself. All things are networked. Everything is interconnected and linked. The nodes of a network are its figures and the interaction among all the nodes or figures with each other are the ground of the network.7 In Part III of this study we generalize hyperlinking to include all processes whereby information is linked from one source of information to another. There are a number of ways in which this process has been described in the literature: linked, interlinked, hyperlinked, networked, internetworked, intertwined, intertwingled, interconnected, interdependent, interwoven, interrelated and associated. The actual network of these interconnections are referred to simply as a network, or as a rhizome, a web, a matrix and in the case of digital networks as the Internet or intranets.

We will explore the way in which one body of information is linked to another and hence we will generalize the notion of digital hypertext that emerged with the World Wide Web to include all forms of linking and hence we will use the term hyperlink to refer to any situation that links two bodies of information. It is a generalization of digital hypertext. Basically, we want to describe the history of hyperlinking. We are stimulated to do this because hypertext has now become an integral part of digital based information. Hypertext and hyperlinks are a way to deal with the information overload of digital media. And since the creators of the Information on the Web interlink their content with the content of others it became apparent to us that interconnectedness is a part of the very nature of what it is to be human in the digital age. Not only are things themselves interconnected but we humans are able to discern that interconnectedness and we imitate it in our speech, writing, and our thoughts. To be alive and aware is to be interconnected and therefore we believe an enquiry into the nature of interconnectedness is an appropriate and important thing to study. Hypertext has brought to our attention the importance of the linking of different texts created in different contexts and not only text-based information but all forms of information in general.

We will focus on and develop an understanding of the interconnectedness of hypertext, hypermedia, hyperlinking and the labyrinth-like structures they create. Hyperlinking is basically the linkage of one body of information with another and as we will show is not limited to digital media but exists in oral and written traditions. A subset of hyperlinking is hypertext in which one body of text is linked to another body of text. Another subset of hyperlinking is hypermedia in which one medium is linked to another medium.

The idea of hypertext came into prominence with digital media because of the ease with which one body of digitally embedded text can be linked to another body of digitally embedded text in the same document or in another document via the Internet. But not only can text be linked to a document but so too can images, audio files and video files. Although the notion of hyperlinking came into prominence with digital media we will show that hyperlinking has been a part of human communication from its very outset and it has been used in oral and written communication long before it came into prominence with digital media.

Chapter 12 will focus on pre-digital forms of hyperlinking and hence the roots of digital hypertext in digital media. Although the term hypertext and the interest in hypertext arose in the context of digital media, it is the case, however that the first forms of hypertext can be traced back to the origin of human communication and its subsequent expression in terms of the spoken word in the recital of epic poetry thousands of years ago and in terms of the written word, first in hand written manuscripts and later in printed documents including books, magazines and newspapers. An example of pre-digital hypertext in the oral tradition that we will consider is the way in which the reciters of epic poetry linked pre-memorized formulae to spontaneously create a narrative that suited the expectations of their audience. We then turn our attention to the linkages within written texts beginning with the marginalia and illuminations found in handwritten manuscripts followed by the footnotes, endnotes, references, indexes and tables of content found in printed books. We will also study the printed form of the Talmud which explicitly makes use of comments on the same page of a central text from Hebrew scriptures. Other examples of the linkages of text in newspapers, magazines, comic books, encyclopedias and dictionary will be examined.

Chapter 13 will focus on the origin of and the impact of hypertext in various digital media including the Internet, the World Wide Web, computers, tablets, and smart phones. This chapter will focus on the hyperlinking between two bodies of text not in the same document.

In Chap. 14 will examine the digital hyperlinking that takes place in the non-textual visual media of movies and television.

Finally, in Chap. 15 of Part III we will examine the nature of the hyperlinking of information in the following five spheres of natural and human phenomena including:

1. the biosphere, the sum of all animate or living beings on our planet Earth;
2. the symbolosphere, the sum of all symbolic representations of human communication, verbal and behavioral and therefore includes language;
3. the sociosphere, a special part of the symbolosphere, is the sum of all social interactions and includes culture;
4. the econosphere, a special part of the symbolosphere and the sociosphere, is the sum of all economic commercial interactions as well as the forms of government that regulate those economic and commercial interactions; and.
5. the technosphere which is also part of the symbolosphere and is the sum of all technological forms.

Chapter 12
Pre-digital Forms of Hypertext

Introduction: The Early History of Hypertexting

The idea of hypertext and hypertexting arose with the Internet and the World Wide Web in which information from one document or Web site is linked to information in another document or Web site automatically when the user clicks on a hypertext link of text that is underlined in blue. If we define hyperlinking or hypertext more generally as the linking of one set of data or information with another set of data or information in another location, it is possible to identify pre-digital forms of hypertext in written documents and books including both handwritten and printed manuscripts or books. We will therefore make use of the definition of N. Katherine Hayles (2001), an American postmodern literary critic and electronic literature theorist more useful because it pertains to both pre-digital and digital hypertext. She wrote, "hypertext has at least these three elements: multiple reading paths; text that is chunked in some way; some kind of linking mechanism that connects the chunks together so as to create multiple reading paths."

We will argue that the recital or performance of epic poetry such as that of Homer entails a form of hyperlinking as defined by Hayles. Another example of pre-digital hypertext as pointed out by Ted Nelson (1965, 1974), who coined the term hypertext, are footnotes and annotations. We shall argue in this chapter that hypertext has had a long and interesting pre-digital history by citing examples including the recitation of epic poetry; the marginalia and illuminations of hand-written manuscripts; as well as the tables of content, the indexes, the footnotes and end notes of printed texts. Although hypertext basically operates in digital media, we have identified a number of examples of pre-digital hypertexting as follows:

- Epic poem of the oral tradition such as Homer as revealed by the work of Parry (1993) and Lord (1960);
- Sacred literature and folk tales in which links are posited between the everyday life of people and the forces of nature in the heavens or skies and the underworld of which they have no control;

R. K. Logan, I. Pruska-Oldenhof, *A Topology of Mind*, Mathematics in Mind,
https://doi.org/10.1007/978-3-030-96436-8_12

- Marginalia and illuminations of hand-written manuscripts;
- Footnotes, endnotes, indexes, tables of content and references of printed books;
- The referencing in articles and books to other articles and books;
- The cross-referencing within encyclopedias, dictionaries and multivolume sets of reference material;
- Shakespeare sonnets and other collections of poetry where a certain theme is visited and revisited;
- The Bible consisting of many different books in which references are made to earlier material inviting the reader to return to these earlier references (the first five books of Hebrew scriptures, the Torah, were composed orally and only committed to writing during the time of the Hebrews exile in Babylon);
- The Talmud with its rabbinical comments and interpretations of portions of Scriptures and the comments on the previous comments of other rabbis. The Talmud was first hand written and later appeared in print;
- Comic books, where the same characters appear over and over again in different issues;
- A series of novels in which the same characters appear like the Harry Potter series of novels;
- Newspapers, where front page stories are continued on back pages and where today's stories refer back to the previous stories that appeared in the same newspaper;
- Magazines where stories are continued on the back pages and where one story interconnects with another story in the same issue of the magazine or in a previous issue of the magazine;
- Academic journals where articles reference each other;
- Modernist literature; and
- The telegraph, telephone and fax in the sense that sets of data or information between non-co-located agents are linked.

In the next chapter we will consider digital hypertext, as Nelson originally imagined it, and as it now exists on our digital devices, on the Internet and in the Cloud. However, in this chapter we will demonstrate how that same structure of linking separate chunks of information using digital hypertext can be found in the pre-digital eras of (1) the oral tradition, (2) handwritten manuscripts, (3) typographic print and (4) electric mass media as outlined above. The linking we have in mind does not always involve a direct link as is the case with end notes, footnotes or indexes where the location of the linked text is explicitly indicated. The linking could be the use of a character from one text that is reintroduced in another text. Examples of this include the link between the same characters in the Harry Potter books or Mark Twain's books Tom Sawyer and Huckleberry Finn. Telephone conversations as well as fax or telegraph communications also represent the linking of separate sets of data or information.

We will also argue that even metaphors represent a kind of hyperlinking in that the same word is used in different contexts, one context is literal and the others are metaphorical. The use of the word cool to denote a temperature between lukewarm and cold is also used to describe a form of jazz, an acceptable or desirable form of behavior, a fashionable item of clothing or as McLuhan used the term to describe a form of media, a cool medium that invites participation as opposed to a hot (another metaphor) medium where the audience is passive. A metaphor links one set of experiences with another. A language, like English, is also about connections or links in that all the words of a spoken language are interconnected and derive their meaning from their interactions with all the other words in that language. A language is a hyperlinked warren of meanings and connections. No word stands by itself. This is evident in the creation of a dictionary, in which every word in the dictionary can only be defined in terms of all the other words in the dictionary. And there are even links to other languages through loan words, such as is the case with English in which one finds the following words: bouquet and chauffeur (French), seltzer, hamburger and wiener (German), kibitz (Yiddish), hombre (Spanish), ciao and pizza (Italian), czar (Russian), modus operandi (Latin) amen and messiah (Hebrew), canoe (Caribbean from *canáoa*), kayak (Inuit), dingo (Australian aborigine). Dictionaries are a labyrinth of meanings. The same holds for encyclopedias that are essentially labyrinths of knowledge. It takes knowledge to create still more knowledge. Every form of knowledge is hyperlinked to all the other forms of knowledge. A library is, in this sense, a hyperlinked collection of books and manuscripts. Every book in the library makes reference to some other book many of which are in the same library if it is big enough. This is certainly the case of the American Library of Congress or an academic library such as the Robart's Library at the University of Toronto.

Catalogues of books and other documents represent a form of hypertexting in that the information in the catalogue links to other collections of information namely the books and documents that are catalogued. The cataloging of library collections dates all the way back to Sumer in the Third Millennium BCE not long after the invention of writing itself by the Sumerians, which means there were forms of hypertexting more than 4000 years ago. At first catalogs made reference to the first line of the scroll or the book being catalogued. Eventually catalogues were organized alphabetically by authors and later by subject. Today's catalogs are now by and large digital with some catalogues providing direct access to electronic versions of a book or an article in an electronic journal.

Early encyclopedias are another example of how information was hyperlinked non-digitally. The model for an encyclopedia was created by Marcus Terentius Varro (116–27 BCE) in a collection of books known as the *Nine Books of Disciplines* each different volume of which was devoted to grammar, rhetoric, logic, arithmetic, geometry, astronomy, musical theory, medicine, and architecture. Pliny the Elder (23–79 AD) was another early encyclopedist and author of Naturalis Historia. Modern day print-based encyclopedias had two levels of hyperlinking through their indexes and by reference in each article to other articles in the encyclopedia. With the emergence of digital encyclopedias, the index and the links between articles are now hypertext links that can be clicked upon.

Oral Hyperlinking

We have just suggested that hypertexting or hyperlinking can be traced back to the very beginning of writing. Actually, we can trace hyperlinking back to the oral tradition before writing if we consider the way in which epic poetry was generated through the use of oral-formulaic composition as was described by Milman Parry (1993) and his student Albert B. Lord (1960), author of *The Singer of Tales*. According to Parry and Lord oral poets or the singer of tales, as in the Homeric tradition for example, composed their epic poems extemporaneously by combining elements from a store of formulae that they had memorized. Each formula had a certain metrical signature (six colon Greek hexameter in the case of Homer) and represented a certain key idea. During the performance of a poem before a live audience the poet or singer of tales would combine or hyperlink these stored formulae to generate a unique story. The hyperlinking took place in the poet's memory that was facilitated by the meter and the rhyme of the stored formulae. The amount of information that could be stored in this manner could be quite extensive. Eric Havelock (1963) in his book *Preface to Plato* describes Homer as a "tribal encyclopedia." He wrote,

> Yet we propose now to look at the poem, so to speak, from the reverse end of the telescope, not as a piece of creative fiction, but as a compilation of inherited lore. Consider then the Muse of the first book of the Iliad as though, while celebrating 'the mighty deeds of former men', she was recording what Hesiod also says she records, namely 'public usage and private habit of all', whether men or gods: as though in fact her utterance did conform to Plato's conception of Homer as a sort of tribal encyclopedia.
>
> We shall deliberately adopt the hypothesis that the tale itself is designed as a kind of convenience, that it is put to use as a kind of literary portmanteau which is to contain a collection of assorted usages, conventions. prescriptions, and procedures' (ibid., 66)

The Iliad and the Odyssey are not just tales of a war and a journey home from a war, but they contain all the information a Greek needed to know to operate properly in their society. Homer's epic poems served as a "tribal encyclopedia" transcribed to memory through the devices of poetry into hyperlinked formulae. They are a compendium of the wisdom of a culture. It should be noted that Homer was not the original composer of the formulae and the Iliad and the Odyssey. Homer is most likely a mythical figure. The date of the composition of the opus attributed to Homer based on the references to geographic locations is well before the emergence of the Greek alphabet. So, what goes as the work of Homer is most likely the transcriptions of the epic poetry compiled into the two collections of the Iliad and the Odyssey.

In addition to the oral tradition of long ago there is another form of oral hyperlinking if one considers the way in which a conversation or dialogue transpires in the everyday affairs of all of us. As the individuals in a conversation or a dialogue take turns they often refer back to things said earlier in the conversation.

Wisdom Lore and Literature and the Metaphor of the Tree: The Tree of Knowledge, the Tree of Life, the World Tree and the Tree of the Knowledge of Good and Evil

Wisdom is interconnectedness to different forms of knowledge, experiences and values that appears in the wisdom lore and literature extant in all societies and cultures. Riffing on Havelock's notion of the "tribal encyclopedia," wisdom lore and literature is a culture or society's encyclopedia. One of the universal metaphors in the wisdom lore and literature of many different cultures is the tree, which to our mind is a symbol of interlinking or hyperlinking of information, experience, knowledge and values to get the big picture so necessary for wisdom. As it is said in the Talmudic Tractate Avot Pirkei (The Wisdom of the Father): "Who is wise? One who learns from every man". And as is stated in Psalms 119:99, "From all my teachers I have grown wise" Wisdom involves combining the knowledge of all and by taking all into account. Wisdom comes from the interlinking of the teachings of many.

The notion of the tree and its association with wisdom because of its many forms of interconnectedness through its branches upwards and its roots downwards is a universal sacred symbol in many cultures across the world. It is in fact, according to Mircea Eliade (1988) for these cultures the backbone of the universe without which the universe would collapse and be destroyed.

> The "closed world," the hollow sphere, has as its center the cosmic pillar separating heaven from earth, the axis which supports the Universe. This symbol of the axis and the pole, the cosmic pillar, is found in all traditional cultures. It is illustrated especially in the civilizations of Mesopotamia, Indo-Melanesia, and Austro-Asia. The "pillar" supports the world; that is, it separates heaven from earth just as does the Egyptian god Shu. This pillar is represented in the "Tree of Life"—all these symbols are equivalent. They support the world, they are the Axis of the Universe, they are the center of the world (Eliade 1988, 137).

The Tree of the Knowledge of Good and Evil

The notion of a tree as a form of interlinking is an ancient symbol associated with wisdom and goes by the following three names The Tree of Knowledge, the Tree of Life and the World Tree. They are even older than the Tree of the Knowledge of Good and Evil (עֵץ הַדַּעַת טוֹב וָרָע or Etz ha-da'at tov va-ra) described in the Genesis story of Adam and Eve in the Garden of Eden found in verses 6–9 of Genesis 2.

> 6: But there went up a mist from the earth, and watered the whole face of the ground. 7: Then the LORD God formed man *of* the dust of the ground, and breathed into his nostrils the breath of life; and man became a living soul. 8: And the LORD God planted a garden eastward in Eden; and there he put the man whom he had formed. 9: And out of the ground made the LORD God to grow every tree that is pleasant to the sight, and good for food; the tree of life also in the midst of the garden, and the tree of knowledge of good and evil (Holy Scriptures 1951).

The Tree of Life also described in Gen 2:9 is connected to the idea of wisdom whereas the tree of the knowledge of good and evil is about morality. The Tree of Life is described in greater detail in Hebrew Scriptures in the book of Proverbs in the following verses:

3:18 She is a tree of life to them that lay hold upon her; and happy is every one that holdeth her fast.

11:30 The fruit of the righteous is a tree of life; and he that is wise winneth souls.

13:12 Hope deferred maketh the heart sick; but a desire fulfilled is a tree of life.

15:4 A soothing tongue is a tree of life, but perverseness therein is a wound to the spirit (ibid.).

The connection between the notion of a tree and wisdom pre-dates the Biblical sources by hundreds of years. There is even a suggestion that the tree of the knowledge of good and evil dates back to post-Akkadian Mesopotamia of 2100 BCE in a seal that depicts two seated figures a man and a woman facing a tree with a snake behind the seated figures flanking them. The seal known as the Temptation Seal or the Adam and Eve Seal was discovered by George Smith (1840–76), an assyriologist who also discovered and translated the Epic of Gilgamesh.

Another archeological find of an even earlier period in Mesopotamia has also been linked to the temptation of Adam and Eve found in Genesis 3:1–7 where the serpent entices Eve to eat of the fruit of Tree of the Knowledge of Good and Evil.

In 1932 E. A. Speiser of the University Museum of Pennsylvania discovered a seal near the bottom of the Tepe Gawra Mound twelve miles from Nineveh. He estimated that the seal came from about 3500 B.C. It shows a naked man and a naked woman, both bent over as if they were oppressed or downcast. Behind them, partly broken, is a serpent (see drawing below in upper left of image). The seal is small, only about one inch in diameter, engraved on stone. (http://www.bible-archaeology.info/adam.htm)

The Metaphor of the Tree, the Tree of Life, the World Tree and the Tree of the Knowledge: An Ancient Symbol of Wisdom and the Interconnectivity of Knowledge

The Tree of Knowledge, the Tree of Life, the World Tree all of which predate the Tree of the Knowledge of Good and Evil are basically interrelated as each is a metaphor for the interconnectivity of the world/universe as well as a metaphor for wisdom. The Knowledge, Life or World Tree is a metaphor for interlinking because it grows on the surface of the earth connecting the sunshine of the heavens to the underworld through its roots. Although they have different names these trees basically incorporate the same notion. They provide a framework whereby the culture that adopted them could understand the world they live in and, as Eliade indicated in the quote above observed, they seem to be a universal theme running through these cultures right across the globe from the New World to the Old World and from Scandinavia and Siberia to India and Southeast Asia among cultures that one would not expect to be linked. Examples can be found in the oral and written folk cultures of the Norse, Germanic folk tales, Hungarians, Greeks, Finns, Balts, Siberians, Turkic folk, Hindus, Chinese folk, Australian Aboriginals, and the indigenous folks of North, Meso- and South America. This need for the interlinking of the elements of one's world parallels hyperlinking.

In the wisdom literature of these peoples, the trees of Life, Knowledge and the World all link to varying degrees to knowledge, wisdom, the origin of life, to their deity or deities and as the Axis Mundi as the structural support of the universe connecting the earth, the underworld and the heavens. The term cosmic tree is often used to classify these mythical trees as we make a world tour. We will make use of this term as we provide the reader with a world tour of the role of cosmic trees in the wisdom lore and literature of the world's peoples in the order listed above.

Norse Legend: According to the Norse legend Edda, Yggdrasil is an immense ash tree that stands at the center of the universe. It is considered holy as the gods assemble there every day to council with each other. Yggdrasil links the heavens through its branches with the underground as it is supported by three roots, one that extend to the spring Hvergelmir the source of all water, one to the well Mímisbrunnr, the source of wisdom that lies under the giant ash tree Yggdrasil. The third root Urðarbrunnr, also lies under Yggdrasil and reaches a lake where the three Norns, Urðr, Verðandi, and Skuld come from. These three female deities determine the fate of both mortals and the gods. Yggdrasil connects all aspect of Norse life.

Hungarian Legend: The universe is divided into three realms connected by the World Tree: the Upper World or the heavens, the foliage of the World Tree is where the gods live and where the good souls go after they die. It is also the home of Isten who controls the fate of the mortals. The Middle World, the trunk of the World Tree is where the mortals live. The underworld, entangled in the roots of the World Tree, is where the evil souls go after they die and are punished.

Finnish and Uralic Legends: These legends are similar to the Hungarian legend with a World Tree connecting the Upper, Middle and Under Worlds.

Latvian, Lithuanian and Old Prussian Legends: These Baltic people had a World Tree with its upper branches reaching the heavens and its roots going underground deep into the Earth.

Meso-American Legends: The legends of the Mayans, Aztecs, Olmecs, Mixtecs and other cultures in Meso-America contain a world tree that connects the heavenly, terrestrial and underground worlds like their Old World Counterparts but with the addition of connecting the four cardinal points of North, East, South and West.

Miscellaneous Legends: The notion of the tree plays a role in many other cultures all over the world. The Tree of Life appears in the religion or legends of the Assyrians; the Ancient Egyptians (in the form of the acacia tree); the Ancient Persians (in the form of mythic Gaokerena tree from which all seeds emerged; the Hindus (in the form of the Eternal Banyan Tree, *Akshaya Vata*); the Buddhists who believe the Buddha attained enlightenment while meditating under the Bodhi tree, a banyan tree; the ancient Chinese who believed the tree of life produces a peach every 3000 years that provides immortality to those that eat it; the Ojibway of North America (in the form of Grandmother Cedar); the Quero Inca believe all trees and plants are sacred and represent their Tree of Life with the Chacana or Incan cross that connects the upper, middle and lower worlds and certain Australian aborigines who believe the Dreaming tree of life and death allows the dead souls to climb the tree to the Sky World.

The Metaphor of the Tree as a Symbol of Interconnectivity in Modern Times

The symbolism of the tree as a way to represent the interconnectedness of information has survived into modern times and has influenced our understanding of life and evolution. Darwin in his *Origin of Species* used a tree structure to describe evolution and show the relationship of different forms of life.

> The affinities of all the beings of the same class have sometimes been represented by a great tree... As buds give rise by growth to fresh buds, and these if vigorous, branch out and overtop on all sides many a feebler branch, so by generation I believe it has been with the great Tree of Life, which fills with its dead and broken branches the crust of the earth, and covers the surface with its ever branching and beautiful ramifications.—Charles Darwin (On the Origin of Species—1859)

Even before Darwin proposed his theory of evolution naturalists such as Augustin Augier, Jean-Baptiste Lamarck, and Edward Hitchcock in the early nineteenth century were using a tree structure to represent the relationship of different forms of life.

The metaphor of a tree is also used in family genealogy in terms of a family tree which links an individual with all their ancestors and all of their family.

In addition to the tree structure some plants have a rhizomic structures as is the case with raspberry plants in which roots spread out underground giving rise to additional raspberry canes and giving the impression of separate plants. Some tree like the banyan tree drop roots from their branches that over time grow into a woody trunk. This allows the tree to grow horizontally like the raspberry canes. With time it is difficult to distinguish these new trunks of the banyan tree from its original trunk. All of these new trunks are interconnected to the original trunk of the tree and to each other. Another interesting facet of the interlinking of plants is described by Beronda Montgomery in an article entitled "Plants thrive in a complex world by communicating, sharing resources and transforming their environments" (https://theconversation.com/plants-thrive-in-a-complex-world-by-communicating-sharing-resources-and-transforming-their-environments-156932, accessed July 18, 2021).

The evolutionary tree that links various forms of life has in addition to a tree structure some horizontal branches and hence a rhizomic structure because of horizontal gene transfer which deviate from the more common vertical gene transfer that is responsible for the basic tree structure of the genealogy of living organisms. This is a topic we will treat in another article that focuses on hyperlinking in the biosphere. Family trees also have a rhizomic structure because of marriages.

The Talmud

The wisdom literature making use of the metaphor of the tree that we referenced in the last section was more part of the oral tradition than a truly literary form. The Talmud in its printed form is a purely literary form that most closely resembles digital hypertext as each page consists of a central text that is surrounded by comments on the central text and often comments on the comments. The Talmud began its life in the oral tradition as the Oral Torah and was passed down generation to generation orally. Its contents were memorized exactly unchanged from generation to generation. Its purpose was to interpret the Torah the basic document of law for the Jewish community. It also provided a set of prescription for the followers of Judaism in terms of the laws of conduct and ritual practices not found in the Torah or Pentateuch, the first five books of the Hebrew Bible. The other books of the Hebrew Bible consist of the 7 books of Nevi'im (Prophets) and the 11 books of Ketuvim (Writings) are not about law. It is only the Torah that contains the 613 commandments (mitzvots), which by tradition are considered to be the Law as handed down to Moses on Mount Sinai "written with the finger of God." In Ex 31:18 one finds: "And he gave to Moses, when he had finished speaking with him on Mount Sinai, the two tablets of the testimony, tablets of stone, written with the finger of God."

With the destruction of the Second Temple circa 70 CE it was decided it was necessary to commit the oral interpretation of the Torah (the Law) to a written document because of the dispersion of the Jewish people into the diaspora. This document became the Mishnah and forms the first part of the Talmud. The Mishnah therefore contains the legal interpretations of the Torah or the Law.

The second part of the Talmud, the Gemara was composed as written text and appeared only after the Mishna or oral interpretation of the Torah was transcribed. The Gemara is an elucidation of the Mishna in which Rabbi's debate the conflicting interpretations of the Mishna as well as discussing other matters that pertain to Jewish life. It contains commentaries on the interpretations of the Mishna and commentaries on commentaries. It is a record of a debate of Rabbi's over time including commentaries found in more recent editions of the Talmud.

The original Talmud was a hand written document in which new commentaries were inserted as appendices. These appendices were a form of hypertext in the sense that the commentaries referred back to the original pages of the Talmud. This all changed with the first printed forms of Talmud. In 1483 only 43 years after the first appearance of the Gutenberg press Joshua Solomon Soncino printed the first individual tractates of the Talmud. Soncino's innovation was not only the use of print, but also the way in which he formatted the pages of the Talmud with the orginal Talmudic text in the middle of the page and the commentaries of two rabbis Rashi and Tosfot surrounding the central text in the margins (see the accompanying figures).

The first printed form of the complete Talmud was published in Venice with the approval of Pope Leo X by Daniel Bomberg in 1519 and then again in 1523 with corrections to the 1519 edition. Bomberg was a Christian who collaborated on the publication of the Talmud with Felix Pratensis, a Jewish convert to Christianity. Bomberg made use of the Soncino's format of a central text surrounded by marginalia. He also added the commentaries of other Rabbi's. His format became the model for all future editions of the Talmud down to this day. It has the same structure of digital hypertext without the actual feature of clicking on a link to take one automatically to the linked text. One merely reads the commentaries surrounding the central text.

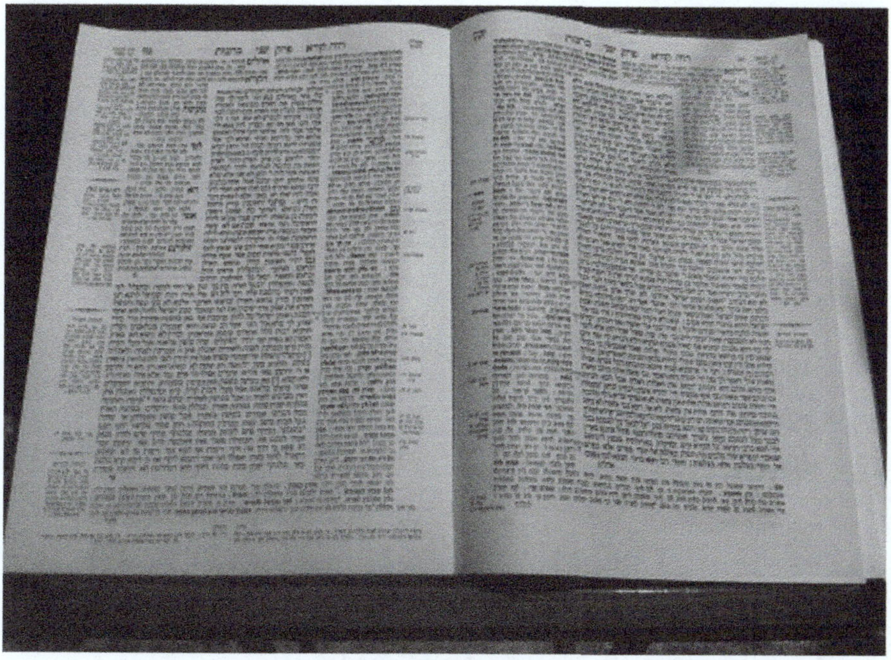

כירה פרק שלישי שבת

The Talmud can be described as a spiral, the oldest texts occupying the center of the page with commentaries spiraling out through the margins across subsequent centuries. French philosopher Emmanuel Levinas (1990, 5) described the Talmud as

> mischievous, laconic...but in love with the possible, [it] register[s] an oral tradition that came to be written down accidentally...in which multiple, but not arbitrary, meanings arise and buzz in each saying. These Talmudic pages seek contradiction and expect of a reader freedom, invention and boldness.

Adin Steinsaltz (1976, 62) calls it a:

> collection of paradoxes: its framework is orderly and logical, every word and term subjected to meticulous editing, completed centuries after the actual work of composition came to an end; yet it is still based on free association, on a harnessing together of diverse ideas reminiscent of the modern stream-of-consciousness novel.

The Talmud is the perfect example of a medium that creates an acoustic space, for it invites active participation from multiple human faculties. Indeed, the Talmudic model of organizing information and knowledge encourages debate and conversation, an active engagement with the work and its past students. Each new edition of the Talmud contains new commentaries so that the Talmud is a living breathing document that continues to grow in complexity. It is a hypertexted dialogue of Jewish thinkers.

Chapter 13
Hypertext in the Digital Domain of Cyberspace

Introduction

Everything is interconnected and linked. No man is an island and by extension no text is an island. Nor does anything or any text stand by itself. Every figure has its ground and operates in some environment. And no environment stands by itself, but is connected to other environments. There is nothing in the literature that stands by itself. Every text digital or non-digital text is connected to other texts, some explicit as in hypertexting and some subliminally.

The focus of this chapter will be on hypertexting in the digital domain of cyberspace having completed our treatment of hypertexting in the non-digital domain in the previous chapter.

Hypertext is defined in many ways but almost always in the context of digital media unlike the way we used the term in the previous chapter. In the Webster-Merriam dictionary, hypertext is defined as "an arrangement of the information in a computer database that allows a user to get information and to go from one document to another by clicking on highlighted words and pictures". The definition found in Wikipedia, the hypertexted open access encyclopedia is:

> **Hypertext** is text displayed on a computer display or other electronic devices with references (hyper[text]links) to other text which the reader can immediately access, or where text can be revealed progressively at multiple levels of detail (https://en.wikipedia.org/wiki/Hypertext).

These two definitions require that hypertext literature be made and experienced on a computer, and is thus medium-specific. Some scholars take a different approach to defining hypertext, one that challenges the digital media-chauvinist argument. In the last chapter we made use of N. Katherine Hayles (2001) definition of hypertext

R. K. Logan, I. Pruska-Oldenhof, *A Topology of Mind*, Mathematics in Mind,
https://doi.org/10.1007/978-3-030-96436-8_13

that has "at least these three elements: multiple reading paths; text that is chunked in some way; some kind of linking mechanism that connects the chunks together so as to create multiple reading paths." We will still make use of this definition in this chapter where we focus on hypertext and hyperlinking associated with digital media with the added refinement that we will use the term 'hypertext link' as the text in a digital document that the user clicks on to link to multiple texts or other bodies of information such as video or audio files. We will use the term hyperlinking to refer to all forms of linking from one body of information to another.

When one hyperlinks data, one creates additional information through the network structures that are created that can have a tree-like, rhizomic or labyrinth-like structure. Hypertext is a special literary form of hyperlinking limited to written or spoken words. Examples of hypertext in digital formats include but are not limited to computer networks, the Internet, the World Wide Web, individual Web sites and Web pages, and Wikipedia to mention a few examples. Hypertext structure is a series of connections with no real beginning or end, just bunches of data to be stumbled upon. The order of stumbling becomes a path, which is rarely in the same order twice (it is difficult to backtrack if you can't see where you've been like Alice in Wonderland, down the rabbit hole).

In this chapter we will examine the structure of hypertext as it was originally developed in Ted Nelson's program Xanadu and its precursors in printed texts and annotations. We argue that hypertext utilizes modular theory to expand nodes of information and organize it to serve a higher function than pure input. These nodes of information diverge and expand into a network of information, with no center and no margins. It is the acoustic space of Marshall McLuhan. We will argue that hypertext is a method of organization that gives the reader/user more agency than non-hyperlinked text. It is by this agency that information is turned into knowledge.

Vannevar Bush and the Memex

The idea of hypertext was first conceived in a certain sense by Vannevar Bush (1945) in an article he wrote for the Atlantic Monthly entitled "As We May Think" where he introduced the idea of the Memex (memory and index) a hypothetical device in which individuals would compress and store all of their books, records, and communications, "mechanized so that it may be consulted with exceeding speed and flexibility" and serve as a device that could hold humanities collective memory. Bush did not think of this as a computer-based system but rather he described the Memex system as an electromechanical device using microfilms enabling individuals to develop and read a large self-contained research library, create and follow associative trails of links and personal annotations, and recall these trails at any time to share them with other researchers. This device would closely mimic the associative processes of the human mind, but it would be gifted with permanent recollection. As Bush writes, "Thus science may implement the ways in which man produces, stores, and consults the record of the race."

The technology used would have been a combination of electromechanical controls, microfilm cameras and readers, all integrated into a large desk. Most of the microfilm library would have been contained within the desk, but the user could add or remove microfilm reels at will. A Memex would hypothetically read and write content on these microfilm reels, using electric photocells to read coded symbols recorded next to individual microfilm frames while the reels spun at high speed, stopping on command. The coded symbols would enable the Memex to index, search, and link content to create and follow associative trails.

Theodor Nelson and Project Xanadu

Theodor Nelson, perhaps inspired by the Memex model, began working on his own program that could link information together and mimic the associative thought patterns of its user. Nelson (1965) coined the terms hypertext to refer to linking bodies of text using electronic computers in his 1965 article *Complex Information Processing: A File Structure for the Complex, the Changing and the Indeterminate*. That term, hypertext has gained common currency in English and most of the languages across the globe. Nelson first formulated his ideas for hypertext when he was a graduate student at Harvard as early as 1960 when he began what he called Project Xanadu. Nelson saw a vision for a "digital repository scheme for world-wide electronic publishing". He proposed a machine-language program that would store and display documents together with the ability to perform edits. This was different from a word processor (which was not invented yet) in that the functionality would have included visual comparisons of different versions of the document, a concept that Nelson would later call "intercomparison". Other elements of Xanadu included non-sequential writing, in which the reader could choose her own path through an electronic document. Nelson called the idea "zippered lists" in a paper he submitted to the Association for Computing Machinery). Zippered lists allowed compound documents to be formed from pieces of other documents, a concept called transclusion. In 1974, with the advent of computer networking, Nelson refined his thoughts about Xanadu into a centralized source of information, calling it a "docuverse". Nelson (1974) published his ideas in his 1974 book *Computer Lib/Dream Machines*. *Computer Lib/Dream Machines* are written in a non-sequential fashion: it is a compilation of Nelson's thoughts about computing, among other topics, in no particular order. It contains two books, printed back to back, to be flipped between. Computer Lib contains Nelson's thoughts on topics which angered him, while *Dream Machines* discusses his hopes for the potential of computers to assist the arts. In his 1982 book *Literary Machines* Nelson (1982) further elaborates on his Xanadu project and other related ideas such as "tumblers," "transclusion" and "micropayments.

Tim Berners-Lee, Hypertext Transfer Protocol and the World Wide Web

Where McLuhan's mind was a spiral, Nelson's was a labyrinth, a network of associations and connections that due to his attention deficit he was unable to connect into a coherent workable system. Nelson formed a company to commercialize his ideas but Project Xanadu never bore fruit. His ideas lived on in the creation of the World Wide Web by Tim Berners-Lee in 1989 which is based on Berners-Lee development of the Hypertext Transfer Protocol (HTTP) which later evolved in HTTPS (Hypertext Transfer Protocol Secure), which provides security over computer networks. HTTPS is the basis for a myriad of Web applications such as Wikipedia and millions of Web sites that allow members of the global village to share and connect data in a hypertextual cyberspace, whose center are everywhere and margins nowhere.

The Internet and the Web emerged to deal with the information overload created by computers and they became the platform for the universal use of hypertext. Hypertext existed before we had digital hypertext as we described in the previous chapter, but because of the ease with which hypertext can be automatically accessed with a click of a mouse, hypertext has become explicit, whereas before Internet it was implicit. With the use of hypertext today via the World Wide Web and other digital platforms that make use of HTTP and patterned on the ideas developed by Bush for his Memex proposal and by Nelson for his Project Xanadu we are experiencing a profound shift in our interconnected culture, one that extends the world over.

Hypertext Is a Labyrinth That Tames and Contains Information Overload

The World Wide Web provides us with an endless interconnection of the world's information and knowledge with access to virtually almost all the information and knowledge that has ever been created. To use Nelson's term the Web provides us access to the "intertwingularity" of human knowledge. Following trails of associations for research (or fun) can lead to hours of information gathering. Attempting to follow the path in reverse at the end of a day on the Web can be difficult if not impossible. In many cases, the links one jumps to in one moment might be erased the next moment. Hypertext readers might find themselves very much like Alice in Wonderland, lost in a maze of paths with no idea how to get home. In this instance, a line of breadcrumbs might be useful, or perhaps a golden string. But what if we did not have that string? Hyperlinks often change colors once a user has clicked on it. Most often, a hyperlink is blue and changes to purple once it has been explored. On Google, however, users can see in history which search results they have already probed, which is very helpful when looking for articles online. Hyperlinks can very easily turn into a labyrinth or a rhizomic network, one with no beginning and no ending, like McLuhan's definition of acoustic space "whose center is everywhere and whose margin is nowhere (McLuhan and Watson 1970, 39)." As a result, users are often desperately trying the same paths over and over, looking for an escape. In this way, hypertext is like a labyrinth originally designed and constructed by Daedalus in Greek mythology to prevent the monster Minotaur from escaping. We, the users of the Web are that Minotaur.

Hypertexts are considered to have a labyrinthine structure. In Greek mythology, the labyrinth was an elaborate structure designed and built by the legendary artificer Daedalus for King Minos of Crete. Its purpose was to hold the Minotaur, a grotesque monster typically described with a head of a bull and the body of a man. The name Minotaur is a portmanteau of Mino for Minos and taur for taurus, the Greek word for bull. The Minotaur was a monster that ate humans and therefore had to be contained and trapped. The labyrinth was so elaborate that the monster was unable to escape from its center and in one version of the story the labyrinth was so complicated that even its builder, Daedalus, had trouble finding his way out of the very labyrinth he constructed.

Labyrinths are a recurring pattern in gardens and the term is often used interchangeably with a "maze". According to Umberto Eco (2014), labyrinths come in three patterns: (1) the classical labyrinth, which is unicursal, that is, one path circling itself; (2) An irrweg labyrinth, (German for "wrong path") which is a branching maze with a single center and finally, (3) the network, where all paths lead to every other path. Each node is connected to every other node, and anyone inside it can travel in every direction, through time and space. The labyrinth links different chambers or nodes to one another. Hypertext is a way of connecting various nodes of knowledge.

According to classical mythology, the minotaur was trapped in the irrweg model of labyrinth. Interestingly, some hyperlinked documents, such as annotated academic writings or fictional texts with footnotes, share this multicursal labyrinth structure because the footnotes lead from the main body of text and cannot stand on their own as narratives. Ultimately, there is one path that leads to the conclusion and that is the main body of text.

Hypertext (and other forms of hyperlinking) is like a labyrinth in that it entails a path that continually changes direction depending on the whim of the reader. Each hypertext link represents a fork in the road where a decision is demanded either to continue in the same direction by ignoring the link or clicking on the link and thereby changing directions and entering a new channel. This parallels the passage through a labyrinth where each fork in the maze demands a decision.

Hypertext and Hyperlinking Have the Structure of a Multicursal Labyrinth

A hyperlinked document has the structure of a multicursal labyrinth because of its non-linear structure in which the reader has a number of choices to make each time they encounter a hypertext link. A unicursal labyrinth is a maze with a tortuous path that leads from an entry point to some final goal. Here are two examples:

Two unicursal labyrinths:

https://commons.wikimedia.org/wiki/File:Cretan-labyrinth-double.svg and https://en.wikipedia.org/wiki/Turf_maze

Here is a comparison of a unicursal labyrinth and a multicursal labyrinth (Fig. 13.1)

https://interlab100.com/literature-review-文献研究/museums-museology/the-labyrinthine-aesthetic-in-contemporary-museum-design/

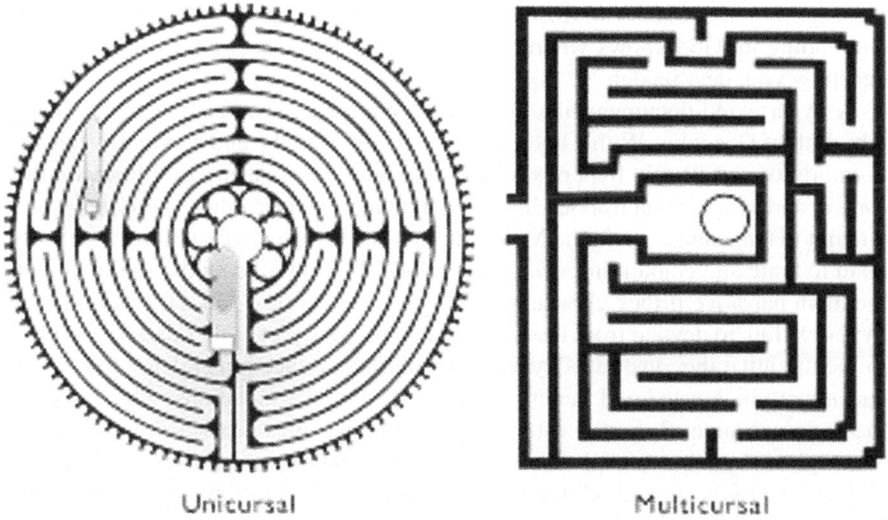

Fig. 13.1 The two paradigms of labyrinth design: the unicursal (left) featuring a single path with no dead ends or bifurcations; the multicursal (right) featuring multiple choices between paths, some leading to dead ends. Paul Basu

A multicursal labyrinth has multiple paths depending on the choices one makes each time the path bifurcates, with some paths leading to dead ends. Some hyperlinked documents have simple diversions where one can pick up additional information but one ultimately returns to the main narrative which has a unique ending. The printed book or article with footnotes or end notes is an example of this kind of structure. It has somewhat the structure of a labyrinth but it is only with digital texts with hypertext links that the narrative assumes the full structure of the multicursal labyrinth. Wikipedia is an example of a text or a document with the full multicursal labyrinth structure.

Wikipedia can be considered a networked labyrinth, as the only beginning it has is when the user starts a session sometimes at the homepage of Wikipedia but more often than not one arrives in Wikipedia as a result of searching for some topic via a search engine like Google. One can even question as to whether or not Wikipedia is actually a document as it does not have the structure of a beginning, a middle and an end. There is no real end as there are always more links one can pursue. The only end is when the reader or user decides to quit. So, in terms of beginning, middle an end Wikipedia is essentially all middle. This corresponds to McLuhan's definition of acoustic space in which, centers are everywhere and margins are nowhere. In other words, what McLuhan called centers are what we referred to as 'middle,' which is everywhere and there is no beginning or end, which are the margins.

The World Wide Web which contains Wikipedia is also without a beginning, middle or end and has a multicursal labyrinth structure. One can start almost anywhere in Wikipedia or on the Web and quit once one has exhausted one's curiosity. One can become easily trapped within Wikipedia's cross linked labyrinth of words moving from one topic to another as has so often happened to us exploring one idea and ending up God knows where. We are sure this has been, most likely, experienced by you, dear reader.

A hypertexted document like the Web or Wikipedia can become a trap like the labyrinth of Deadalus. With a book, the reader comes to the end and that's it there is no more to read. But in hypertexted environments, the reader can follow the links forever and ever and get trapped like the Minotaur. The Deadalus's labyrinth was an attempt to tame and contain the Minotaur. In a similar way hypertexting is a mechanism to tame and contain information overload. It allows the reader to remain with the main narrative if they choose not to click in a hypertext link or by clicking the hypertext link to deviate from the main narrative and explore other narratives related to the narrative they just left. Hypertexting allows connections related to the main narrative to be created and thereby building a web of associations, which enriches the main narrative. But as so often happens one becomes distracted and abandons the original narrative to explore a new one, which has become more compelling.

Hypertext is a method of structuring data and information we come across over history and in our daily lives in order to hold off information overload, in much the same way that Daedalus built the labyrinth to hold the Minotaur in the myth. In extending this metaphor, information overload is the Minotaur, the violent and unnatural creature which Dante includes as being the guard to the seventh level of Hell and which we have built structures, grammar, rules and infrastructure around to stave off.

The labyrinth links different chambers and levels of this chaos so humans can navigate information freely without being overloaded. Just as the labyrinth of Deadalus entraps the Minotaur so hyperlinks entrap the attention of the reader or listener. Hypertext is a way of connecting various nodes of knowledge.

As civilizations evolve, more and more information is uncovered, requiring ever more complicated mazes to keep the Minotaur trapped. With the quickening speed of communication and travel and the fragmentation of reality that ensued during the Industrial Revolution, modernist writers such as James Joyce in Finnegan's Wake, Ezra Pound in Cantos and T.S. Eliot in the Wasteland found ways of mirroring this fractured psyche in their works. As a response to the fractured mentality of modernity, exemplified by works such as Nabokov's *Pale Fire*, post-modern writers used semiotic theoretical techniques of leaving ambiguous endings and no narrative closure, as well as academia-inspired footnotes and endnotes, typically used for citation purposes, as a way to restructure a narrative.

Often considered one of the most difficult works of fiction, James Joyce's Finnegan's Wake is another example of ergodic literature. Umberto Eco called it an "open text" in that it can be interpreted an infinite number of ways. However, it is not a hypertext. Due to the structure of the first and last line, it is considered a closed narrative, in that the last line links to the first and the narrative closes in on itself like an Einsteinian universe. There are no alternative endings or paths in the printed version Finnegan's Wake, although the text itself can be interpreted ad infinitum. Therefore, it is an example of a unicursal text, in Aarseth's terms. Finnegan's Wake in printed form is not a hypertext but its openness lends itself easily to the hypertextual format. (See http://finwake.com/ for an example of Finnegan's Wake as hypertext).

David Foster Wallace, known for his persuasive arguments for the use of unconventional, but nonetheless logical, alterations to narrative form, began using footnotes in his journalism, essays and even novels. In an email sent to his editor as an explanation for his use of footnotes in his book Infinite Jest, Wallace argues that they allow him:

> To make the primary-text an easier read while at once 1) allowing a discursive, authorial intrusive style w/o Finneganizing the story, 2) mimic the information-flood and data-triage I expect'd be an even bigger part of US life 15 years hence. 3) have a lot more technical/ medical verisimilitude 4) allow/make the reader go literally physically "back and forth" in a way that perhaps cutely mimics some of the story's thematic concerns … 5) feel emotionally like I'm satisfying your request for compression of text without sacrificing enormous amounts of stuff (Max 2009).

Wallace's email to his editor describes the forethought of the structure of literature, the piecing together of narrative information, the author's intentions, and the emphasis on process and materiality of writing and reading, that is, experiencing, a book. Through reference to James Joyce's Finnegan's Wake, an experimental novel notorious for its convoluted time lines and often unfinished trains of thought, Wallace sought to use footnotes as a way to organize his narrative, in a way that a typical narrative within only one body of text was unable to accomplish. Wallace also points out how the reader's experience of the book would change, giving them the choice to "literally physically 'back and forth'" with the text, and to be aware of its materiality and formal existence. Using Aarseth's understanding of how a footnote is both unicursal and multicursal, in that it gives the reader the option to choose various reading methods, David Foster Wallace's use of footnotes falls into Hayles' (2001) description of hypertext (it must have multiple reading paths, text that is chunked in some way, and some kind of linking mechanism). Footnotes and/or endnotes contribute to all these categories, and seems to fulfill the need to order fragmented narrative that is attuned to postmodern writers. Wallace believed "that reality is fractured right now… The difficulty about writing about that reality is that text is very linear, it's very unified. I am constantly on the lookout for ways to fracture the text that aren't totally disoriented. You can make the lines and jumble them up, and that's nicely fractured, but nobody's going to read it" (David Foster Wallace interviewed by Charlie Rose, March 27, 1997, see item 13 on https://www.amazon.com/Charlie-Rose-March-1997/dp/B00301ZLOQ).

When Marie Mundaca designed the interior of the collection of essays, *Consider the Lobster* (Wallace 2005), she helped Wallace capture the type of fractured reality he had sought elsewhere to capture with footnotes and endnotes. "Wallace's idea was to have leaders and labels, like a diagram," she said. "He wanted something that looked like hypertext rollovers that were immediate and at hand (Mundaca 2009)." His essays took on new form and new life, with pieces of disjointed information poking out of the main text, tantalizing the reader to break up there linear reading habits and embrace their fragmented reality, by reading a text that mimics the "constant monologue inside your own head (Wallace, Kenyon Commencement Speech,

2005, web.ics.purdue.edu/~drkelly/DFWKenyonAddress2005.pdf, accessed March 24, 2017)." In many ways, David Foster Wallace's fractured narrative style fits right in to the definition of what people commonly consider hypertext on a computing device, instead he used old forms inspired by academic modes to push the boundaries of print-based text, using footnotes and blocks of text within the text, metatext, to exfoliate the narrative.

Now, much of Wallace's work has been digitized, with footnotes made into hyperlinks and sidebars for the reader to click on should she choose. Some might say this is the perfect medium for Wallace's work, an environment that makes "the primary text easier to read" and certainly mimics the information data-triage of the present day, but this certainly is not the only way to read Wallace. As pre-existing forms get sucked in to various mediascapes, users and creators alike feel more confused, lost and uncertain than they were with print. Faced with the World Wide Web, and multitudes of online hypertext environments, fracturing and linking narratives is becoming more and more common. While authors have a new medium to play with, some still struggle with what to include in the content. Many find themselves turning to past stories to bring with them on the journey of digitalization. In doing so, they undergo transformation as an artist and that the content undergoes a transformation as well.

Alice's adventures in wonderland have become a prominent metaphor for traversing new media. This is in part due to the mediated journey the story itself took, from oral storytelling on a boat to printed manuscript to cartoon to Imax and beyond. Another factor in Alice's reemergence is the story's absurdity. It is easy to take a nonsensical fable and imprint new meanings to it in almost any capacity. Perhaps it is pure convenience that we use this metaphor now to understand the world through hyperlinks. And yet there are parallels between the remediation of Alice's story and the remediation of human knowledge. Stories told in oral culture were eventually transcribed in manuscript form, passed along through monasteries and churches and once the printing press arrived new interpretations and new knowledge began proliferating the world. So, has Alice been reinvented and reimagined with new technologies. Consider the 1991 movie The Matrix. This film made use of the Alice metaphor to lure Neo out of the Matrix and into a world where he could see clearly the effect of the media around him.

We are not proposing that this book will shed light on an apocalyptic world, we would like to shed light on the effects of interconnectedness and network theory that has and will continue to proliferate human lives. And so, upon embarking on our journey to research seemingly simple hypertexts, we wound up falling down the rabbit hole of network theory, labyrinths and hyperlinks, and becoming quite intertwined in the process.

It comes as no surprise that Alice falling down the rabbit hole is often used as a metaphor for exploring new media environments and styles. The original book has many allusions and references to philosophical concepts, theories on logic, mathematics, and even rhetoric. Yet the passive reader might not understand or notice all these extra-textual references—ideas, notes and theories that exist outside the book. So, *The Annotated Alice* (Caroll and Gardner 1960) came into existence, combining

Alice in Wonderland and the sequel, Through the Looking Glass, filled with endnotes to point the reader in the direction of further reading. Now, though, while many people have heard the story of Alice, very few have read the book. So, to connect the story to as wide an audience as possible, Alice in Wonderland has been remediated, translated into several different digital copies, including a copy on Project Gutenberg, where readers have to click links to read the next chapter.

While many theorists, Aarseth (1997) and Carr (2008), for example, argue that clicking hyperlinks promotes active engagement with the text, clicking the screen to read from one chapter to another seems just as passive, if not more so, than flipping pages in a book. As Aarseth tactfully explains, "the emerging new media technologies are not important in themselves, nor as alternatives to older media, but should be studied for what they can tell us about the principles and evolution of human communications" (Ergodic Literature). Translating older texts that were meant for printed books into what is known now as hypertext, from "HyperText Markup Language", does not show that the hypertext structure is better suited to digital media, though it certainly is reaching new possibilities and wider audiences. Instead, it shows that the form is universal, a means of exploring human communication through various forms, media, and languages, digital or otherwise.

The narrative forms that authors such as James Joyce, Vladimir Nabokov and David Foster Wallace were experimenting with is merely testing out the waters of a new environment. As N. Katherine Hayles (2001) writes, "it would be a mistake … to think that hypertext is only an electronic form, for an exuberant tradition of print hypertext exists … where nonlinear narrative pathways are the rule rather than the exception" We should celebrate, then, the materiality of such "works" with multiple pathways, the design and structure that goes in to making these printed books differ from tradition, linear narrative forms. Hypertexts are not limited to one medium or the other, we should instead consider the form as a genre, one that cannot be contained by its physical attributes. If hypertext was inspired by such experimental literature as Nabokov's Pale Fire (Gary Wolf 2015), surely the two media, print and digital, can still learn much from each other.

Hypertexted Cyberspace and the Structure of McLuhan's Acoustic Space

Despite the use of hypertext in our World Wide Web or maybe because of it, cyberspace looks remarkably similar to the chaos of Borges' (1941) library as related in his short story The Library of Babel published in 1941 in his collection of stories in *The Garden of the Forking Paths*. The information found on the Internet has its centers everywhere and its margins nowhere, which is precisely McLuhan's definition of acoustic space as evidenced by the following excerpts from his writings:

> Acoustic space is totally discontinuous, like touch. It is a sphere without centers or margins as Prof. Bott of the University of Toronto explained a generation ago (McLuhan 1970, 194).

> The world of acoustic space whose center is everywhere and whose margin is nowhere, like the pun (McLuhan and Watson 1970, 39).

> Acoustic Space has the basic character of a sphere whose focus or center is simultaneously everywhere and whose margin is nowhere... Acoustic space is dynamic; it has no fixed boundaries. It is space created by the method or process itself. In contrast, visual space is static and container-like, with a fixed center and margin (McLuhan and Powers 1989, 74).

When being interviewed by Playboy Magazine McLuhan was asked, "What do you mean by acoustic space?" He responded,

> I mean space that has no center and no margin, unlike strictly linear space, which is an extension and intensification of the eye. Acoustic space is organic and integral, perceived through the simultaneous interplay of all the senses; whereas 'rational' or pictorial space is uniform, sequential and continuous and creates a closed world (McLuhan 1969).

By way of an interesting sidebar we want to share with the reader that McLuhan's definition of acoustic space is almost identical to a twelfth century definition of God (Findlay-White and Logan 2016): "God is the intellectually knowable sphere whose center is everywhere and whose circumference is nowhere" This definition was used by many theologians, philosophers and one librarian. This list includes Alain de Lille, Meister Eckhart, Nicholas of Cusa, Giordano Bruno, Robert Fludd, Blaise Pascal, Gottfried Wilhelm Leibniz and Jorge Luis Borges.

In acoustic space, data and information come at you from all directions at once. It is simultaneous and instantaneous. Visual space, on the other hand, is sequential and linear. McLuhan claims before writing, ideographic and alphabetic were invented and before there was print media, the world was acoustic. Oral poetry was the popular culture of the time. Nothing was set in stone, and so information and stories would change depending on the context in which they were experienced. The epics of Homer belong in the acoustic tradition. When writing was developed, the world was plunged into a visual world. History was written on a linear, two-dimensional plane. Information was regarded as true if it stood the test of time. But it was subject to being destroyed. Information was not privy to change. With the advent of electric information, such as the telephone, radio and television, McLuhan claimed that society was entering a new acoustic age. Although he did not live to see the emergence of the digital age and world of hypertext he somehow foretold of its arrival.

Not only did McLuhan brilliantly describe the revolution that electricity and electric mass media had created but he also prophesized and described the next revolution of digital media, a revolution he never had a chance to observe.

Arthur Porter, a colleague, related how McLuhan who never saw a personal computer during his lifetime predicted its arrival in a 1968 luncheon with a group of IBM executives:

> Mac Hillock, [an IBMer] arranged a lunch with half a dozen of IBM's divisional directors. Marshall got soon tuned up and was telling them about a computer for every home, no need to visit the grocery store... Two of them said to me after lunch, "we have not heard of anything as crazy as that!" Marshall was talking about the personal computer a dozen years before they thought of it. Here was a professor of English more than a decade ahead of the

> technical people in computer evolution. He was thinking in terms of the user (Nevitt and McLuhan 1994, 29–30).

McLuhan foreshadowed the hyperlinking found on the Internet, with services like Google and Amazon:

> Instead of going out and buying a packaged book of which there have been five thousand copies printed, you will go to the telephone [read Google and/or Amazon via the Net], describe your interests, your need, your problems, …and they say it will be right over. And they at once Xerox, with the help of computers from libraries of the world, all the latest material just for you personally… They send you a package as a direct personal service. This is where we are heading under electronic information conditions. Products increasingly are becoming services (McLuhan et al. 2003, 99).

McLuhan (1967, 67) seems to have described the Internet in the following passage from a 1967 article:

> The computer in education is in a very tentative state but it does represent basically speeded up access to information and when it is applied to the telephone and to Xerox it permits access to the libraries of the world, almost immediately, without delay. And so, the immediate effect of the computer is to pull up the walls of the subjects and divisions of knowledge in favor of over-all field, total awareness—Gestalt.

If one substitutes packet switching for the telephone and if one read Xeroxing as the use of a printer, then McLuhan's description in the above quote is an accurate description of the Internet written two years before the arrival of the ARPANET, the forerunner of the Internet. In 1962 he wrote:

> A computer as a research and communication instrument could enhance retrieval, obsolesce mass library organization, retrieve individual encyclopedic function and flip into a private line to speedily tailored data of a saleable kind (http://en.wikipedia.org/wiki/Marshall_McLuhan).

One finds in these lines a foreshadowing of Wikipedia as the retrieval of "individual encyclopedic function" and the foreshadowing of Google as the generator of "saleable" data in the way Google generates advertising revenues from its searches.

Encyclopedias and dictionaries have been some of the most crucial methods of information storing and consulting since the invention of movable type and the printing press and they continue that tradition with digital media and the use of hyperlinking with their online versions such as Wikipedia and the online version of the Encyclopedia Britannica (www.britannica.com), which no longer prints a hard copy version of their encyclopedia. An online version of a dictionary or an encyclopedia is so much more convenient that a hard copy version precisely because of the ease of navigating using hypertext.

Online dictionaries and encyclopedias resemble the associative patterns of human cognition by using bifurcation and set theories to allow for an infinite number of possible combinations of information within the finite structure of the online Web site of the dictionary or encyclopedia as links to other sites enrich what the online dictionary or encyclopedia has to offer. They use techniques such as hypertext to bifurcate the traditional visual and linear space of print and thereby provide the reader with more agency. This makes the online encyclopedia a "cool medium" as defined by McLuhan because the user, by interacting with the text, participates in creating their own version of the encyclopedia by following the links that their curiosity suggests to them. Printed encyclopedias were cool media to a certain extent through their use of end notes and footnotes and their references to other articles in the encyclopedia but they were nowhere as interactive as is the case with online encyclopedias. This illustrates McLuhan's insight that "the user is the content" by which he meant that the reader/user through their interpretation contribute to the content of what the author has composed. With hypertext the reader determines the content by the hypertext links they choose to click on.

Ergodic Literature

If online dictionaries and encyclopedia provide more interactivity that their print versions the same is true of a new class of literature that also make use of hypertext links that actively engages the reader and allows them to help the author create a narrative that appeals to their interests or predispositions.

Texts can appear as any form of a labyrinth. The classic structure is the simplest, where the reader travels through the narrative from start to finish, accepting the path the author carves out for them. An irrweg text is more complex. The author leaves multiple paths for the reader, but ultimately there is one outcome. This form of text is called "ergodic literature,"

a term coined by Espen J. Aarseth (1997) in his study Cybertext-Perspectives on Ergodic Literature. Ergodic derives from the Greek words "ergon" meaning work, and "hodos" meaning path. This type of reading compels the reader down multiple pathways while giving them the choice to fill in the text with extra nodes of information. Typical narrative print books, Aarseth argues, requires trivial amounts of activity and engagement, because the reader merely turns the page and moves their eyes back and forth. Ergodic literature, on the other hand, requires non-trivial engagement, because the reader/user must click on links and follow various paths to follow the narrative.

In his dissertation, E.J. Maloney explains that paratext, that is, footnotes and endnotes, "frequently play a significant role in the progression of narrative fiction, often extending the boundaries of the narrative frame, introducing new heuristic models for interpretation, and offering alternative narrative threads for the reader to unravel" (Maloney 2005). Since footnotes/endnotes in fiction give readers the choice to read the text passively, straight through, or actively, by jumping from the main text to the notes—it can be said that printed books can in fact be examples of ergodic literature. Indeed, Aarseth himself explains that, "the footnote is a typical example of a structure that can be seen as both uni- and multi-cursal. It creates a bivium, or choice of expansion, but should we decide to take this path (reading the footnote), the footnote itself returns us to the main track immediately afterward. Perhaps a footnoted text can be described a multicursal on the micro level and unicursal on the macro level" (Aarseth 8). Thus, printed fiction with footnotes and diegetic paratext can have the same properties as digital born literature, the former using footnotes as the linking structure to send the reader to different nodes of text, and the latter using hyperlinks to the same effect.

Aarseth continues to make his point through reference of the old metaphor for the structure of narrative, the labyrinth. The most basic difference between a standard narrative and a hypertextual narrative is that the former resembles a unicursal maze (beginning from one definite starting point and following it through to a single end), while the latter takes on a multicursal structure (with many different beginnings and endings, and various routes and paths that have no clear ending at all). Aarseth brings up Nabokov's Pale Fire, a printed poem which includes fictional paratext and annotations, as an example of a text that "can be read either unicursally, straight through, or multicursally, jumping between the comments and the poem" (8). *Pale Fire* is not only an example of printed ergodic literature; it is also considered a precursor to digital hypertext format. It was printed one year after Ted Nelson coined the term hypertext. Nelson even asked for permission to use Pale Fire in a demonstration of his hypertext system Xanadu. Though this collaboration never happened, it is easy to see the connection between printed experiments with narrative forms and the hypertext systems of today.

Ergodic literature would also describe what Marshall McLuhan called a "cool medium". The network is most complex of all. In a text with no center and no end, the possibilities are infinite. An example of this kind of text is Wikipedia, or any online crowd-sourced article, where multiple authors can change the body of text at will and change the outcome of the text. Hypertext can only occur at the last two labyrinth patterns, since a unicursal text has only one pathway and therefore no branching nodes of text. The unicursal text is the epitome of linear reading.

Labyrinths are a recurring pattern on coins and gardens and the term is often used interchangeably with a "maze". According to Umberto Eco (2014), labyrinths come in three patterns: (1) the classical labyrinth, which is unicursal, that is, one path circling itself; (2) An irrweg labyrinth, (German for "wrong path") which is a branching maze with a single center and finally, (3) the network.

A unicursal labyrinth is a maze with a tortuous path that leads from an entry point to some final goal. A multicursal labyrinth has multiple paths depending on the choices one makes each time the path bifurcates, with some paths leading to dead ends. Some hyperlinked documents have simple diversions where one can pick up additional information but one ultimately returns to the main narrative which has a unique ending.

The third labyrinth pattern is the most complex. A network with no center and no margins like Wikipedia. There are starting points, presumably, but these are infinitesimal, meaning that there are an infinite variety of entry points into this grand document. In this kind of labyrinth, all paths lead to every other path. Each node is connected to every other node, and anyone inside it can travel in every direction, through time and space.

Information is not static in a network. It changes depending on the way it is reached and the context in which the user/reader discovers it. Wikipedia is not a static encyclopedia. Every hour of the day, someone somewhere is editing an article or adding new information. It is a collaborative effort by all the world's knowledge creators with access to the Internet. Wikipedia uses associative structure through hyperlinking. If ever a reader wants to learn more about a specific topic mentioned in an article, she merely must click on the hyperlinked word or phrase and she is brought to a new section of information. If, by chance the word she wants is not hyperlinked, she has the option to gather information and create her own Wikipedia article. Or not. In a sense, hypertext replaces footnotes and has the advantage that it gives immediate access to the material contained in a footnote. In essence, hyperlinking is a way of organizing information. The author, or an outside source, fragments the main narrative of the text, giving the reader multiple reading paths. The reader is trapped in a labyrinth of information and attempts to follow a specific path to get to the end.

The ways we read hypertext—non-linear, discursive patterns—can change the way society understands the development of history. The image of the network, the most complex of the labyrinth patterns, obsolesces the classic unidirectional labyrinth as well as the traditional tree-like pattern. In *From the Tree to the Labyrinth* (Eco 2007), Umberto Eco calls for a rearranging of the way information is organized. Rather than the tree structure, where every node of information branches off from a central focal point, Eco argues that a web structure, which is wholly infinite, is better suited for structuring information. As McLuhan put it, the Global Village is a spiral, who's center is everywhere and margins nowhere. Perhaps the metanarratives so despised by the poststructuralists were in fact the main focal point of the Tree of Knowledge. Following the network structure of the World Wide Web and extending it in order to reimagine the structure of society's language, culture and biology (rhizomes).

Chapter 14
Hypermedia: Hyperlinking Cinema and Television

The Origin of Hypertext and Hypermedia

We have used the term hyperlink to describe any linking from a body of information in one medium to a body of information in that medium or possibly in another medium. Examples include hyperlinks from a Web-based text file to another Web site, a video, a podcast, etc.

In Chap. 12 we extended the notion hyperlinking to include non-digital links of text and oral recitation. In this chapter we will examine non-digital hyperlinking in cinema and television. By the term hypermedium or hypermedia we are referring to a medium or media in which the content is not exclusively text and hence our consideration of hyper linking in cinema and television, that is what we will call hypercinema and hypertelevision (see https://en.wikipedia.org/wiki/Hyperlink_cinema).

Hypermedia is another term developed by Ted Nelson (1965). An extension of the term hypertext, hypermedia refers to any medium that is non-linear that also includes graphics, audio, visuals, plain text and/or hyperlinks. This differs from multimedia, which may include non-interactive, linear media as well as hypermedia. The World Wide Web is a classic example of hypermedia, whereas standard non-interactive cinema is an example of multimedia in that it combines audio and visual display but does not include hyperlinks. However, there are instances where cinema can be considered a hypermedium.

Cinema is in a unique position to narrate and comment on the diachronic and synchronic connections amongst people, places and things. In a world that seems to be becoming more complex each day, network narratives mirror this same confusion and expansion of human consciousness, responding to and commenting on the quickened pace of a hypermodern world, a theory that has proliferated the social sciences since the 1970s.

Film critic Alison Quart (2005) coined the phrase "hyperlink cinema" to refer to films that that feature the interconnectedness or intertwingularity (Nelson 1974) of characters, multiple time and story lines in the form of flashbacks or flash-forwards,

R. K. Logan, I. Pruska-Oldenhof, *A Topology of Mind*, Mathematics in Mind,
https://doi.org/10.1007/978-3-030-96436-8_14

and split screens or graphic intertitles that connect characters through time and space (Quart 2005). Some examples of Hyperlink Cinema include: *Pulp Fiction* (1994), *The Opposite of Sex* (1998), *Magnolia* (1999), *Crash* (2004), *Happy Endings* (2005), *Babel* (2006) and a host of others. The process of editing the film, through simple cuts, to flashbacks, to graphic intertitles link characters and locations the same way a hyperlink links information.

According to Quart (2005) the film Happy Endings, directed by Don Roos, challenges the notion that the new media of the day is "dumbing us down". Instead, the film seems to claim that multi-task entertainment actually makes us sharper. The film mirrors the hyperlinked cyberspace that society is accustomed too, although it is more likely that this genre is popular due to the shortening attention span than the new sharp-wittedness of society. Complete with captions and split-screen gimmicks, the film uses text and image to link the characters through time and space. These footnotes and extradiegetic texts are more than just "postmodern pompoms", they are extensions of the audience's "mental processing". In a strong director-audience relationship, Roos speaks directly to the audience's narrative desire. It is precisely the film's, and the genre's, natural reflexivity that speaks to human nature in the twenty-first century. In a world full of mechanical extensions of some human faculty, society is naturally self-reflexive.

It has been suggested that hyperlink cinema, or network narratives, "is a mechanism that grapples with the developments of globalisation" (Beal and Minton 2009). However, despite the focus on multiple character connections, studies show that most hyperlink films do not challenge the typical human brain as much as Roos hoped. Based on a study conducted by Krems and Dunbar (2013) films such as Happy Endings still feature an appropriate number of character connections per the average number of social connections in the real world. The study concluded that "audiences cannot track dramatized events that deviate significantly from the constraints of everyday psychology". Even the films that do exhibit a larger social network than average social situations, the richness of the connection between characters is more constrained, meaning there are likely fewer main characters. This is reflective in real world situations as well, where it seems genuinely difficult to host a conversation with more than four people. Despite directors' best efforts, hyperlink cinema seems unable to break through the constraints imposed by evolved human psychology. However, these films reflect more than our ability (or lack thereof) to gather a larger friend list. These films offer a condensed view of the reality of our world, replete with on-screen commentary and information available in an instant.

Alison Quart concludes with a final axiom: "The narrative cleverness of [Happy Endings] may borrow something from popular technology. But [the film] nevertheless gets at something very human, traversing from one state of unfulfillable yearning to another". While new technology and media offer multiple choices and pathways, each with new meaning, there is still no way of knowing which, if any, are the better option. We are, for better or worse, going our journey alone, blindly searching for the path that will get us beyond our current reality.

An extension to hyperlink cinema has been proposed by Robert K. Logan and Emma Findlay-White (Wikipedia 2016) to suggest that hyperlink television has many of the properties of hyperlink cinema in that it also plays with time and characters' personal history, plot twists, contains interwoven storylines between multiple characters, jumps between the beginning and end (flashback and flashforward), and makes use of split screens and footnotes. But in addition, it has some marked differences, to wit: weekly dramatic episodes often with multiple seasons such as *Downton Abbey*, *Boardwalk Empire*, *The Sopranos* and *Game of Thrones*.

Each episode is composed of intertwined short scenes that last no more than a couple of minutes. This is particularly true of the final season of Boardwalk Empire, where Nucky Thomson's childhood and young adulthood scenes are interspersed with his later or contemporary life. Characters such as the Commodore, the Commodore's mistress, Ely, Nucky's brother are featured in the flashbacks. The insertion of historic figures in Boardwalk Empire such as President Harding, Andrew Mellon, Eddy Cantor, Al Capone, Lucky Luciano, Meyer Lanky, Arnold Rothman, J. Edgar Hoover all add to the hyperlinked experience. The use of the contemporary music of the times is another hyperlinked element in Boardwalk. The fact that these series play out in weekly episodes and stretch over many seasons (years) provides another diachronic element compared to a movie.

Game of Thrones is another great example of hyperlinked television, because times, places and characters are edited together to create a linked universe. The information revealed in each episode ties the world together and gathers information for the audience, who eagerly awaits the conclusion to which the show is building.

Another example of hyperlinked television are the nightly newscasts such as the major networks news programs in the USA, the CBC National in Canada and the cable news outlets like CNN, Fox New, MSNBC and in Canada CBC News Network. CNN is perhaps the most iconic form of such as the CNN's *The Situation Room with Wolf Blitzer*. The show consists of two sets of characters that make up the content of CNN. There are those whose activities are reported upon or the newsmakers such as Barack Obama, Hillary Clinton, Donald Trump, Joe Biden, Bernie Sanders, Bibi Netanyahu and a host of interviewees including members of Congress and celebrities. The other set are the anchors, reporters and commentators, and interviewees such as Wolf Blitzer, Anderson Cooper, and Fareed Zakaria. Split-screen expert panels visually link information among people and places, often with an anchor acting as facilitator or game-master. Graphic sidebars and scrolling text fills the screen with even more information, giving audiences information from various newsworthy items that are happening simultaneously.

Video games are also a form of hypermedia, because they promote active participation among users with the content. Many video games have hyperlinked elements whereby pressing a button, standing next to a character or performing an action can unlock new information about the game. Users can call up maps on the screen, inventories of items gathered and a variety of other information to help them through the game. This is typical of single-player games, where the game is mostly plot-driven, as opposed to multi-player games where the content is user-based.

Chapter 15
Hyperlinking in the Spheres: The Physiosphere, Biosphere, Technosphere, Sociosphere and Econosphere

Introduction

In Part III of Topology of Mind, we have focused on hyperlinking in: (1) pre-digital oral and written media (Chap. 12); (2) digital media (Chap. 13) and (3) hypermedia such as Hypercinema and Hypertelevision (Chap. 14). In this chapter we will examine non-human forms of hyperlinking that occur or exist in nature. One can divide the subjects of human study into six nested spheres of interactivity consisting of:

1. the physiosphere, the set of all abiotic matter,
2. the biosphere, the set of all living organisms, that is embedded in the physiosphere,
3. the symbolosphere, the set of all symbolic forms created by humans including language, culture, social and economic systems and technology. Within this sphere one finds the following embedded spheres: the technosphere, econosphere and the sociosphere,
4. the technosphere, the set of all media technologies, tools and human artifacts,
5. the econosphere, the set of all economic and commercial transactions including government, and
6. the sociosphere, the set of all social interactions including culture.

The hyperlinking of information in each of these six spheres will be entertained in this chapter.

The motivation for addressing hyperlinking in the biosphere, technosphere, econosphere, and sociosphere is that these domains have a major impact on the human mind and hence are of interest for any study of mathematics in mind. From our discussion to this point it is obvious that the human mind is not just what happens in the brain or for that matter the brain and the rest of the human body. The mind is extended by the tools we use (Clark and Chalmers 1998; Clark 2003) and by our spoken language and the various forms that speech evolved into (Logan 1997, 2007). In the summer of 1997, Logan (1997) introduced the term the Extended

R. K. Logan, I. Pruska-Oldenhof, *A Topology of Mind*, Mathematics in Mind,
https://doi.org/10.1007/978-3-030-96436-8_15

Mind and the notion that spoken language extended the brain that had hitherto had been merely a percept processor into the mind capable of conceptual thought and planning. Andy Clark and David Chalmers (1998) a few months later in January of 1998, quite independently, introduced their famous Extended Mind Hypothesis that proposed that objects within the environment, primarily tools, function as a part of the mind. They wrote (ibid.),

> Where does the mind stop and the rest of the world begin? The question invites two standard replies. Some accept the demarcations of skin and skull, and say that what is outside the body is outside the mind. Others are impressed by arguments suggesting that the meaning of our words "just ain't in the head", and hold that this externalism about meaning carries over into an externalism about mind. We propose to pursue a third position. We advocate a very different sort of externalism: an *active externalism*, based on the active role of the environment in driving cognitive processes.

The external elements that drive cognitive processes for Clark and Chalmers are basically tools. Their more general hypothesis subsumes Logan's (1997, 2007) idea as language and the various forms it evolved into of writing, mathematics, science, computing and the Internet are basically tools. Andy Clark (2003) elaborated on the Extended Mind Hypothesis in his book Natural-Born Cyborgs: Minds, Technologies, and the Future of Human Intelligence and Logan (2007) elaborated on his hypothesis in his book *The Extended Mind: The Emergence of Language, the Human Mind and Culture.*

Given that there is a biological component to our mind, a technological component because of our tools, a social component through our interactions with other minds and an economic component through our commercial interactions a study of mathematics in mind must take these domains or factors into account. This is the motivation for the study of the linkages and hyperlinkages of in the biosphere, technosphere, econosphere, and sociosphere because each of these spheres influence the mind and hence mathematics in mind.

The Physiosphere and the Absence of Hyperlinking

As previously mentioned, the notion of information does not pertain to abiotic matter as non-living matter makes no choices. Making use of Terrence Deacon's (2012, 524) definition that "information is about something for something towards some end," we can say that information is that which allows a living organism, i.e. a sentient being, to carry out its metabolism, to survive and to reproduce and it is about something for something and hence involves some level of choice. It is only with life (i.e. living organisms) that the notion of information has any meaning.

Some physicist like to talk about the information associated with abiotic matter and their interactions, but they are using the term information metaphorically. These scientists are basically talking about the information they have about the abiotic matter that they study as a result of their observations. As abiotic matter cannot make any choices, they have no need for and hence no interaction with information.

Information can be embedded in abiotic matter as words printed in a book but those signs in the book do not become information unless they are read and interpreted by a reader. The physical book is just the medium by which the author communicates information to a reader. Information as we have pointed out entails the formation of a mind or an intelligence. When the earth and the sun interact gravitationally there is no intelligence and no mind to be informed. Neither the Earth nor the Sun have any intelligence and hence no mind that can be informed. Therefore, it is only in the biosphere that information can exist and have any meaning.

Information allows living organisms human and non-human all the way down to the level of single cell organisms and viruses to make decisions. Abiotic matter makes no decisions as they have no choice to speak of and hence information in the context of their interactions is meaningless. Living organisms are constantly making decisions based on the information they are able to interpret. With this in mind (pun unintended) we now turn to the biosphere.

Hyperlinking in the Biosphere

In this section we will be talking about the hyperlinking of information in the biosphere. We ask the reader to understand that when we represent the many different exchanges of information within the biosphere as hyperlinking we are talking metaphorically. Our motivation is to demonstrate in this study the universality of hyperlinking, which is not just about the exchange and linking of symbolic information among humans in digital, electric, written and oral environments but also the linking and exchange of information in other spheres of operation such as the biosphere, the econosphere, the technosphere and the sociosphere.

The definition of information we will make use of for the biosphere is the one that was developed in the paper, "Propagating Organization: An Enquiry" (Kauffman et al. 2007), and which was called biotic or instructional information. In that paper the notion of the relativity of information was introduced and it was pointed out that there are many different definitions of information depending on the context in which it operates.

> The reader may legitimately ask the question "isn't information just information?", i.e., an invariant like the speed of light. Our response to this question is no, and to then clarify what seems arbitrary about the definition of information. Instructional or biotic information is a useful definition for biotic systems just as Shannon information was useful for telecommunication channel engineering, and Kolmogorov (Shiryayev 1993) information was useful for the study of information compression with respect to Turing machines. The definition of information is relative and depends on the context in which it is to be considered. There appears to be no such thing as absolute information that is an invariant that applies to all circumstances. Just as Shannon defined information in such a way as to understand the engineering of telecommunication channels, our definition of instructional or biotic information best describes the interaction and evolution of biological systems and the propagation of organization. Information is a tool and as such it comes in different forms.

Our definition of information differs from that of physicists in that we believe information is that which informs a living organism to make the decisions that are essential to their survival and that of their off springs. Also, it is worth noting that what is commonly called information on a computer or in a book are really just signals and do not become information until those signals are interpreted by a reader or a user. Shannon's so-called Information Theory is really just a theory of signals.

We will consider in this section the many different ways that information is exchanged back and forth in the biosphere including the exchanges between organisms or within organisms including the following:

- organisms with each other, both among conspecifics and non-conspecifics;
- among conspecifics organisms either as mate selection, social cooperation or competition;
- among non-conspecific organisms where the interaction could be cooperation as in mutualism and symbiosis or it could be antagonistic as is predator and prey relationships;
- all living organisms and the information in their physical environment (i.e. their respective umwelt) that they must perceive and interpret ranging from the simplest single cell bacteria (prokaryotes) to various multicellular organisms (eukaryotes) including vertebrates, mammals and humans;
- the individual organs and cells within all eukaryotes where the channels of information exchange can be through the circulatory system, the nervous system or the immune system;
- the organelles and the nucleus within living cells of eukaryote organisms;
- the exchange of genetic material through vertical gene transfer (VGT) between organisms and their descendants;
- the exchange of genetic material through horizontal gene transfer (HGT also known as lateral gene transfer—LGT) between different species.

The latter two mechanisms, VGT and HGT, for the hyperlinking of information give rise to the evolution of life and the notion of the tree of life, which also possess some rhizomic structure due to HGT. If VGT and natural selection were the only mechanism for the evolution of life one could claim that there is a Tree of Life and that there exists a Last Universal Common Ancestor. With HGT one can still make a similar claim but not all the branches are those of a tree but rather some of the branches are rhizomic where HGT took place.

As we will be considering the hyperlinking of information within the biosphere we will often be talking about the exchange of biotic or instructional information as it was defined in the article, "Propagating Organization: An Enquiry" (ibid.):

> We believe Schrödinger was deeply correct, and that the proper and deep understanding of his intuition is precisely that an aperiodic solid crystal can contain a wide variety of micro-constraints, or micro boundary conditions, that help cause a wide variety of different specific events to happen in the cell or organism. Therefore we starkly identify information, which we here call "instructional information" or "biotic information," not with Shannon, but with constraints or boundary conditions. The amount of information will be related to the diversity of constraints and the diversity of processes that they can partially cause to

> occur. By taking this step, we embed the concept of information in the ongoing processes of the biosphere, for they are causally relevant to that which happens in the unfolding of the biosphere.
>
> We therefore conclude that constraints are information and, as we argue below, information is constraints which we term as instructional or biotic information to distinguish it from Shannon information. We use the term "instructional information" because of the instructional function this information performs and we sometimes call it "biotic information" because this is the domain it acts in, as opposed to human telecommunication or computer information systems where Shannon information operates. This step, identifying information as constraint or boundary condition, is perhaps the central step in our analysis. We believe it applies in the unfolding biosphere and the evolving universe, expanding and cooling and breaking symmetries.

In addition to talking about biotic or instructional information we will also be talking about the perceptions of organisms of their environment as well as the signals they exchange with each other. Each organism has its own umwelt, which are those aspects of their environment which they are aware of and able to perceive. The same environment in which two different organisms operate will result in a unique umwelt for each organism. There are signals that might be information for one organism but not the other and vice versa.

The modes of perception of living organisms are the ways in which they hyperlink to the information in their umwelt, information that is essential for their survival including their access to water, sunlight, nutrition and the location of their prey or predators depending on whether they are the predator or the prey and for some species they can be either one.

Another important aspect of hyperlinking is the communication among organisms, which like the perceptions of their physical environment are also percept based with the exception of humans who are also capable of symbolic communications.

Hyperlinking in Animals

Animal communication entails the transfer of information from one animal to another and can involve conspecifics or animals of another species. Animals communicate information to each other through a variety of channels including auditory sounds such as barks, growls, screeches, songs (as with birds and whales); displays or gestures (for example the peacocks feathers or a dog showings its teeth to indicate its willingness to attack), even by distinctive coloration (such as a wasp or bee communicating do not tread on me); odors by which an animal can signal its willingness to mate or as a warning to a prey animal of the presence of a predator; touch such as grooming or cuddling and through chemical signaling such as the communication networks in social insects like ants, termites and bees that use pheromones as hyperlinks. Communication serves many different functions such as the dance of honey bees signaling the location of food for members of their hive; mating rituals;

alarms alerting conspecifics to the presence of a predator (vervet monkeys, for example, have a vocabulary of four alarm signals depending on the type of predator that is present); staking out territory for food acquisition or for mating; and signaling a willingness to fight.

Examples of the hyperlinking or communication takes place in the way social insects interact with each other to create what is called a super-organism like a bee hive, a wasp nest or an ant colony. Among vertebrae there are many species that are constantly hyperlinking information such as wolves that form packs, cattle operating in herds, fish swimming in schools, birds flying in flocks, elephants travelling in packs, lions hunting in prides, beaver building dams and the list goes on and on. Even bacteria communicating with each other through chemical signaling.

There is also in addition to the hyperlinking of information between organisms the hyperlinking of information within an organism between its various components such as between its internal organs, between its individual cells and within individual cells between the cell's nucleus and its organelles. The channels of communication within an organism include its circulatory system, its nervous system and its immune system.

Hyperlinking in Plants, Fungi, and Microorganisms

In addition to the hyperlinking of information in animals there are also similar processes in plants, fungi, unicellular eukaryotes and bacteria. Plants have the ability to sense when they are under attack by predators and they respond by releasing volatiles, one group warns other plants in their neighborhood of the presence of the herbivore predators and the other group attracts parasites that will attack the herbivores. Plants exchange information with other organisms such as fungi and unicellular organisms with whom they have a symbiotic relationship.

Fungi communicate information with each other making use of pheromones for mating purposes. They also enter into symbiotic relationships with plants, unicellular eukaryotes and bacteria which entail the exchange of information.

Both prokaryote and eukaryote unicellular organisms communicate with each other. In certain species of bacteria through a process known as quorum sensing individual bacteria are able to determine the density of their colony and when that density reaches are certain critical value a new level of gene expression kicks in resulting in one of these three forms of behavior of the colony: biofilm formation, virulence or antibody resistance. Slime molds are a class of eukaryote unicellular organisms that live as single cell organisms when food is plentiful. When this is not the case they communicate and come together to form a slime that can detect the location of food chemically and then move towards that source of nourishment. Slime molds will also form a fruiting body that releases thousands of spores that allows the slime mold to reproduce.

Intercellular and Intracellular Signaling Within a Single Organism

In addition to the hyperlinking of information between interspecies and intraspecies organisms there is also the hyperlinking of information within a single organism due to intercellular and intracellular signaling.

Intercellular signaling occurs when the receptor cell detects a signal sent from another cell. Cells have receptor proteins on their outer membrane with which the molecular signal known as a ligand can bind to. The receptor protein then transmits the information into the interior of the cell which sets off a series of intracellular signals known as transduction that cause the cell to perform its function. This process may or may not involve gene expression in the nucleus of the cell. These processes of signaling are basically nature's form of hyperlinking one body of information to another body of information that allows organisms and their cells to carry out their metabolism.

> **Signal transduction** refers to the transmission of a molecular signal in the form of a chemical modification by recruitment of protein complexes along a pathway that ultimately triggers a biochemical event in the cell. Signal transduction occurs when an extracellular signaling molecule activates a specific receptor located on the cell surface or inside the cell. In turn, this receptor triggers a biochemical chain of events inside the cell—known as a signaling cascade—that eventually elicits a response. Depending on the cell, the response may alter the cell's metabolism, shape, gene generation, or ability to divide. The signal can be amplified at any step; thus, one signaling molecule can generate a response involving hundreds to millions of molecules. Signal transduction involves the binding of extracellular *signaling molecules,* also called ligands, to receptors that trigger events inside the cell (https://en.wikipedia.org/wiki/Signal_transduction).

In the biosphere, the basic levels of nature are hyperlinked. Cells of the body are hyperlinked in the sense that information is passed from one cell to another through various circulatory systems and even within the cell information is linked between the nucleus and the cell's various organelles. For example, the mitochondria (or chloroplast) cannot function without the information from the nucleus and the nucleus with information from the mitochondria (or chloroplast). At the biological level, there are two structures of information: The tree of life structure, as theorized by Darwin, and the rhizomatic structure, as theorized by Deleuze and Guattari.

In arborescent (hierarchic, tree-like) models, information bifurcated from a single source which contrasts with the rhizome structure that resists this type of organization. Gilles Deleuze and Felix Guattari (1987) developed a theory of philosophy and knowledge organization based on rhizomic patterns found in nature. Plants such as strawberries and ginger grow roots on a horizontal plane, which then produce their own roots and nodes. If the root is separated from the plant, it will continue to grow and create its own plant. Deleuze and Guattari extend this image into a metaphor for the organization of knowledge and culture. In their sense, history and culture, much like a rhizome, is a series of connections and links with no specific origin, for a "rhizome has no beginning or end; it is always in the middle, between things, interbeing, intermezzo" (ibid., 7). The planar movement of the rhizome resists hierarchical, top-down methods of organization, instead favoring a nomadic network of casual interconnections.

Hyperlinking in the Technosphere

The technosphere consists of all the tools, technologies, media and artifacts that have been constructed by humankind. We do not distinguish between tools, technologies, media and artifacts. They are just different terms for the same things. A tool is a technology, an artifact and a medium all at the same time. We also assert that all tools are hybrids composed of simpler tools and hence all forms of technology are hyper-linked through hybridization in the technosphere. The only exceptions are the very first tools used by humankind, which were objects found in nature like rocks and sticks that were employed to facilitate some human function. Following the ancient Greeks, we regard nature as everything but humankind and their tools or artifacts. We ask the reader to understand that when we represent the many different exchanges of information within the technosphere as hyperlinking we are talking metaphorically as we did in our treatment of the biosphere. The human agents that invent, manufacture and use their tools facilitate the hyperlinking of tools by combining different simpler tools to create more complex tools.

Let us describe what we mean when we claim that all tools are hybrids. Let's begin with the hand axe used to cuts things. It is a hybrid of the rock from which it was manufactured plus the process of flaking that was employed to create a sharp edge. Another simple tool was a lever which was a hybrid of a stick found in nature and a rock used to create a fulcrum. The hatchet was created by tying a hand axe to a stick and is therefore a hybrid tool composed of a hand axe and a stick with the stick acting as a lever. Other examples of simple hybrid tools include the bow and arrow, and the spear. As technologies became more complex they were built up from many different simpler tools over the millennia that humankind created new technologies until we arrive at today's complex technologies like the automobile and the computer. Both of these modern devices are composed of hundreds of simpler technologies. This is why we claim that technologies in the technosphere are composed of hyperlinked tools. But how do we justify the hybridization of tools as hyperlinking? The answer is simple. Every tool, every technology is a form of the organization of something material and as such it represents a form of information (not the material out of which the tool is constructed, but the design of the tool, the concept behind it). We therefore claim that the hybridizing tools is equivalent to the component tools being hyperlinked, i.e. connecting one body of information associated with one tool with another body of information associated with the other tool.

Hybridization or the hyperlinking of tools is inevitable as McLuhan pointed out when he wrote:

> A new medium is never an addition to an old one, nor does it leave the old one in peace. It never ceases to oppress the older media until it finds new shapes and positions for them (McLuhan and Zingrone 1995).

The hybridization of technologies or "the hybrid or the meeting of two media is a moment of truth and revelation from which new form is born (McLuhan 1964, 55)". Not only is a new form created but also, as McLuhan noted, new forms of force and energy are released: "The crossings or hybridizations of the media release great new force and energy as by fission or fusion (Essential McLuhan, 174)."

When tools are hyperlinked, new technologies are forged which then create or release the new "force and energy" that McLuhan talked about. One can extend this idea to the hypertexting of text and the hyperlinking in the biosphere where new forces and energies are released in these domains. A hypertexted document in which bodies of text are hyperlinked reveal new insights compared to exactly the same text batched in a linear and sequential way. Eukaryotes can be thought of as hybrids of prokaryotes that release new forms of energy and force. Similarly, vertebrates can be thought of as hybrids of invertebrates plus the additional biological feature of a back bone that gives the class of vertebrates their very name.

Hyperlinking in the Econosphere

The sociosphere and the econosphere are closely linked because of the fact that humans are intensely social animals whose survivals depends on their mutual cooperation with each other in which the exchange of goods and services is the key to their survival. In hunting and gathering societies banishment was almost equivalent to a death sentence, since it would be very difficult for a lone human to survive in the bush. If we define the sociosphere as the sum of all social interactions among humans then the econosphere, the sum of all of the economic transactions among humans, is clearly a part of the sociosphere, as an economic transaction is also a social interaction. We will focus first in this section on the econosphere in which the social interactions are economic and in a separate section return to social interactions which are not strictly commercial in nature.

The econosphere is the hyperlinks of all economic or commercial transactions. Commercial transactions include the buying and selling of goods and services, employment, all business transactions, and all business organizations including incorporated and unincorporated as well for-profit and not-for-profit organizations. In this section we will be talking about the hyperlinking of information in the econosphere. Once again, we ask the reader to understand that when we represent the many different exchanges of information within the econosphere as hyperlinking we are talking metaphorically as we did in our treatment of the biosphere and technosphere.

Another way of looking at commerce and economic transactions is that at its roots it represents the exchange of value and is a form of cooperation, symbiosis or mutualism in which both parties gain something of value. This is not to suggest that all transactions or exchanges are fair and equal, but they do involve the exchange of value. And in a certain sense value represents a form of information. Therefore, we claim that commerce that involves the exchange of value also represents the hyperlinking of values and hence the hyperlinking of information.

In addition to the hyperlinking of value there is also the hyperlinking of information within a business organization; between business organizations; and between a business organization and their customers and their suppliers. In addition to these exchanges of information there is also the marketplace that connects consumers and producers, so we also must take into account the hyperlinking of consumers and producers with and within marketplaces.

The beginning of commerce among humans, if we include our hominid ancestors, who also belonged to the genus Homo, was the nuclear family consisting of mother, father and their non-adult children. They lived together as an economic unit in which they helped each other to survive. The nuclear family has always been the basic economic unit even after more complex economic arrangements emerged. With the control of fire nuclear families tended to live with other nuclear families with whom they had some kind of kinship relationship in order to share the benefits of the campfire, which needed the constant attention of members of the clan. Those that hunted or gathered shared with those that tended the fire and, in this situation, we have the first beginnings of commerce. There was an exchange of value but no one was counting. This form of commerce persisted through most of the Paleolithic period with the occasional exchange of goods when different clans or tribes encountered each other in the course of their wanderings. An object of abundance in one's territory might be exchanged for an object that was rare there. With the Neolithic revolution that was labor intensive and in which surpluses could be easily accumulated the exchange of goods and services experience a quantum leap forward. At first commerce was strictly a straight exchange of one product or service for another. To facilitate these exchanges markets were created where those that wished to trade could bring their products or services to a central place at a certain given time to facilitate the exchanges of goods and services. With the population explosion that agriculture created, states emerged to regulate commerce and the medium of money came into use to facilitate complex commercial transactions. Money became a way to measure value and hence became a medium of exchange and a form of information. Money or monetary value acted as a hyperlink between different products and services.

Douglas Rushkoff (2016) in his book *Throwing Rocks at the Google Bus* described how money became hyperlinked to the value of goods and services in the Middle Ages. The economy of peasants and craftsmen operating in the late Middle Ages made use of market or local currencies such as grain receipts or receipts for goods and services rendered. This system of local currencies allowed the farmers and craftsman, the emerging middle class, to prosper at the expense of the aristocracy and the monarch who had hitherto controlled all commercial activity. It was at this stage just as the first stirrings of the Renaissance were occurring that the monarchs instituted two regulations that stifled the emerging middle class's burgeoning economy. The first was the introduction of centralized currency or coin, the coin of the realm, and the outlawing of local currencies. Money has two functions: the storage of wealth (coin of the realm) and the facilitation of commerce and the exchange of goods and services (which local currencies has previously facilitated).

The second constraint on commerce was the establishment of monopolistic chartered trading companies by order of the King. The King was given a share of the company in an exchange for preventing others to compete with the chartered company. This was the beginning of business people bribing the rulers who ran the government of the realm. These two developments represented a change of the operating system of the economy. The medieval economy based on the barter and exchange of the bazaar were replaced by an economy based on a centralized currency based on the precious metals of gold and silver and dominated by monopolistic corporations chartered or licensed by the crown. This created a hardship for farmers and craftsmen who were required to borrow money and pay interest to conduct their commerce and as a result they no longer prospered as before. It was the first round of the 1% versus the 99%. Because craftsmen were no longer able to sell the goods they created in order to survive they had to sell their labor and thereby only realize a fraction of the value they created. They became wage slaves.

With the industrial revolution the chartered corporations were slowly replaced by another form of monopoly, namely industrial corporations capitalized by the wealthy as shareholders to whom the management of the company had to answer. The objective of these companies was growth. Applying McLuhan's "*Laws of the Media* (McLuhan and McLuhan 1988)," Rushkoff showed that these corporations enhanced the power of capital, obsolesced the bazaar or local economies, retrieved empires and flipped into personhood, where under the law corporations were given the same rights as individuals but not the same responsibilities. Rushkoff points out because of their mandate for growth the corporations looked for new places to extract wealth, which in time led to colonialism.

With the establishment of companies, a new form of hyperlinking emerged in which members of the business organization needed to hyperlink their information in order for the business to operate as a wealth creating machine. The hyperlinking of information between the different departments or divisions of a company are essential to its success. The marketing arm of the company had to depend on information from sales and sales was dependent on the creators/designers of products and services. And the creators/designers are dependent on research and development. And so on down the line. A company is like a living organism in which the hyperlinking of information is essential for its survival and its ability to thrive.

The field of knowledge management emerged to facilitate the hyperlinking of knowledge and information in companies that have grown too big for its employees to know each other on a day to day basis. The basic lesson of knowledge management practitioners is that the top-down hierarchical flow of information in a linear sequential one item at a time is inefficient. The basic remedy that knowledge management prescribes is the hyperlinking of information along functional lines that short circuits the chain of command or command and control architecture of traditional business structures.

Hyperlinking in the Sociosphere

As already mentioned one of the elements of the sociosphere is the econosphere, which we have just treated. We now turn to the other elements of the sociosphere often referred to as culture.

Culture is an important adaptive mental tool whereby information is hyperlinked among the members of a society and the learning of previous generations are passed on to the next generation through communication and social interactions. As suggested by Boyd and Richerson (1985, 14):

> Individual learning... can be costly and prone to errors. Learning trials occupy time and energy that could be allocated to other components of fitness, and may entail a considerable risk to the individual as well. Because of these costs, the investments of individuals in determining the locally favored behavior must be limited, and individual learning can lead to errors. Individuals may fail to discover an adaptive behavior, or a maladaptive one maybe retained because it was reinforced by chance. When these costs are important, selection ought to favor shortcuts to learning—ways that an organism can achieve phenotypic flexibility without paying the full cost of learning. Cultural inheritance is adaptive because it is such a shortcut. If the locally adaptive behavior is more common than other behaviors, imitation provides an inexpensive way to acquire it.

Geertz (1973, 8) defines culture as "an historically transmitted pattern of meanings embodied in symbols, a system of inherited conceptions expressed in symbolic forms by means of which men communicate, perpetuate and develop their knowledge about and attitudes towards life."

Whether through imitation as Boyd and Richerson suggested or by communication as Geertz suggested cultural patterns are basically hyperlinked between the members of society in a myriad of ways and channels.

It should also be noted that within any given society there are many different co-existing sub-cultures. Families each have their own sub-cultures as do all organizations within the society such as businesses; governments agencies; religions including individual churches, mosques, temples and synagogues; schools; universities and their various departments; social clubs; sports teams, etc. Various forms of hyperlinking exist within various sub-cultures and between sub-cultures.

Epilogue: Closing the Spiral Loop: The Eternal Return of This Study

In the conclusion we will close our spiral loop and return to where we began.

As we wrote in the opening chapter of this book we have covered many diverse topics related in varying degrees to mathematics in mind. We have examined mathematics in mind from the perspective of the spiral and hyperlinked structures of the human mind in terms of its language, its thoughts and its various modes of communication especially those of artists as in Chapter 11 of this study. What we have discovered along the way as we pursued this study is that the human mind is not a linear deductive logic machine although at times it deploys linear deductive logic. But rather it is a holistic, cyclic, hyperlinked, spiral, and mathematical creative entity operating in acoustic space whose center is everywhere and its margins nowhere linking percepts and concepts, creating and destroying ideas in the arts and sciences, creating and enjoying beauty, loving and caring for others. And still it is a mystery with many of its operations only barely understood such as its ability to be creative, altruistic, and even conscious of its own actions.

Thank you for sharing your mind with ours and feel free to share your thoughts that we may have provoked by emailing us at logan@physics.utoronto.ca and i2pruska@ryerson.ca.

R. K. Logan, I. Pruska-Oldenhof, *A Topology of Mind*, Mathematics in Mind,
https://doi.org/10.1007/978-3-030-96436-8

Epilogue: Closing the Spiral Loop: The Eternal Return of This Study

References

Albright, W.F. 1957. *From the Stone Age to Christianity*. 2d ed. Garden City, N.Y.: Doubleday Anchor Books.

Aarseth, Espen J. 1997. *Cybertext: Perspectives on Ergodic Literature*. Baltimore: Johns Hopkins University Press.

Baker, Kenneth. 2005. "Talking with Robert Smithson." In Lynne Cooke & Karen Kelly (eds) *Robert Smithson: Spiral Jetty*. Berkeley: University of California Press.

Bandle, O., H. Klingenberg, and H. Mauer. 1958. Kategorien Lehre vom Satz. Hamburg.

Bateson, Gregory. 1973. *Steps to an Ecology of Mind*. St. Albans: Paladin Frogmore.

Beal, Wesley and Viv Minton. 2009. "Theorising Connectivity: The Form and Ideology of the Network Narrative." In Fiona Hackney (ed) *Networks of Design*: *The conference proceedings of the 2008 Design History Society Conference*. Boca Raton FL: Universal Publications.

Bloomfield, Leonard. 1933. Language. New York: Holt.

Boyce, Mary 1975. *A History of Zoroastrianism: Volume I. The Early Period.* Leiden: Brill Publishers.

Boyd, R. and P. H. Richerson. 1985. *Culture and the Evolutionary Process*. Chicago: University of Chicago Press.

Boyer, C. B. 1991. *A History of Mathematics (2nd edition).* New York: Wiley

Brakhage, Stan. 1982. "Test of Time." (https://www.fredcamper.com/Brakhage/TestofTime.html, accessed July 30, 2019).

Bush, Vannevar. 1945. "As We May Think." *The Atlantic* [Boston] July 1945: 101–08.

Caroll, Lewis, and Martin Gardner. 1960. *The Annotated Alice*. Bramhall House, 1960. Print.

Carr, Nicholas. 2008. "Is Google Making Us Stupid?" *The Atlantic*. 1 July 2008. Web. 9 Apr. 2015.

Christiansen, Morten. 1994. Infinite languages finite minds: Connectionism, learning and linguistic structure. Unpublished doctoral dissertation, Centre for Cognitive Studies, University of Edinburgh UK.

Clark, Andy and David Chalmers. 1998. "The Extended Mind." *Analysis* 58: 7–19.

Clark, Andy. 2003. *Natural-Born Cyborgs:* Minds, Technologies, and the Future of Human Intelligence. New York: Oxford University Press.

Cook, Theodore Andrea. 2015. *Spirals in Nature and Art: A Study of Spiral Formations*. San Francisco: Andesite Press (originally published in 1903 and reissued by Andesite).

Cook, Theodore. 2009. Spirals In Nature And Art: A Study Of Spiral Formations Based On The Manuscripts Of Leonardo Da Vinci (1903) Whitefish MT: Kessinger Pub Co.

R. K. Logan, I. Pruska-Oldenhof, *A Topology of Mind*, Mathematics in Mind,
https://doi.org/10.1007/978-3-030-96436-8

Cubitt, Sean. 2005. "Media, History of." New Dictionary of the History of Ideas. 2005. Retrieved August 05, 2015 from Encyclopedia.com: http://www.encyclopedia.com/doc/1G2-3424300462.html

Curtis, James (1987) *Rock Eras: Interpretations of Music and Society, 1954–1984*. Popular Press.

Darwin, Charles. 1871. *Descent of Man*. Accesses on Feb 22, 2017 at https://books.google.com/books?id=8iMTCgAAQBAJ&pg=PA92&lpg=PA92&dq="A+complex+train+of+thought+can+no+more+be+carried+on+without+the+aid+of+words,+whether+spoken+or+silent,+than+a+long+calculation+without+the+use+of+figures+or+algebra."&source=bl&ots=t01iKSm1is&sig=2TdjjmK67-T9jjveu77ttbmzj8k&hl=en&sa=X&ved=0ahUKEwix6cu2gaTSAhUs2oMKHfkFCDQQ6AEINzAH#v=onepage&q="A%20complex%20train%20of%20thought%20can%20no%20more%20be%20carried%20on%20without%20the%20aid%20of%20words%2C%20whether%20spoken%20or%20silent%2C%20than%20a%20long%20calculation%20without%20the%20use%20of%20figures%20or%20algebra."&f=false.

Deacon, Terrence. 1997. *The Symbolic Species: The Co-Evolution of the Brain and Language*. New York: W. W. Norton & Co.

Deacon, Terrence. 2012. *Incomplete Nature: How Mind Emerged from Matter*. New York: W. W. Norton and Company.

De Cruz, H. and J. De Smedt. 2015. A Natural History of Natural Theology. Cambridge MA: MIT Press.

Deleuze, Gilles, and Félix Guattari. 1987. *A Thousand Plateaus: Capitalism and Schizophrenia*. Minneapolis: U of Minnesota, 1987.

Donald, Merlin. 1991. *Origins of the modern mind*. Cambridge, MA: Harvard University Press.

Donald, Merlin. 1998. "Mimesis and the executive suite." In: Hurford, J.; Studdert-Kennedy, M.; and Knight C. (eds), *Approaches to the evolution of language* 44–67. Cambridge: Cambridge University Press.

Drake, Stillman. 1957. *Discoveries and Opinions of Galileo*. New York: Doubleday & Company.

De Heinzelin de Braucourt, J. 1962. "Ishango." *Scientific American*, 206–6, June, 105–116.

Eco, Umberto. 2014. *From the Tree to the Labyrinth: Historical Studies on the Sign and Interpretation*. Trans. Anthony Oldcorn. Cambridge, MA: Harvard University Press.

Eco, Umberto. 2007. From the Tree to the Labyrinth. Cambridge MA: Harvard University Press.

Eco, Umberto. 1989. *The Open Work*. Cambridge, MA: Harvard University Press.

Edward, Brooks. 2013. *The Philosophy of Arithmetic as Developed from the Three Fundamental Processes of Synthesis, Analysis and Comparison, Containing Also a History of Arithmetic*. London: HardPress Publishing.

Eliade, Mircea. 1964. *Cosmos and History: The Myth of the Eternal Return*. New York: Harper Torchbooks. Translated from the French by Willard R. Trask

Eliade, Mircea. 1988. *Symbolism, the Sacred and the Arts*. New York: Crossroad.

Eliade, Mircea. 2006. In Bryan Rennie (ed) Mircea Eliade: A Critical Reader. London: Equinox Publishing.

Eliade, Mircea. 1987. *The Encyclopedia of Religion. Vol. 15*. New York: MacMillan Publishing.

Everson, Ted. 2007. *The Gene: A Historical Perspective*. Westport CN: Greenwood Press.

Farber, Zac. "'Neurotic and Obsessive' but 'Not Too Intransigent or Defensive': Editing David Foster Wallace." (<http://www.thehowlingfantods.com/Farber_EditingDFW.pdf>, accessed April 15, 2019.)

Findlay-White, Emma and Robert K. Logan. 2016 "Acoustic Space, Marshall McLuhan and Links to Medieval Philosophers and Beyond: Center Everywhere and Margin Nowhere." *Philosophies* 2016, 1(2), 162–169. Also available online at http://www.mdpi.com/2409-9287/1/2/162.

Freiberg, J. 1981. "Methods and traditions of Babylonian mathematics. Plimpton 322, Pythagorean triples, and the Babylonian triangle parameter equations." *Historia Mathematica* 8, 1981, 277–318.

Galilei, Galileo. 1623. *Opere Il Saggiatore* (The Assayer). Rome. Translate by Drake (1957, 237–38).

Geertz, Clifford. 1973. *The Interpretation of Culture*. New York: Basic.

Grosswiler, Paul 1996. "The Dialectical Methods of Marshall McLuhan, Marxism, and Critical Theory". *Canadian Journal of Communication* 21 (1).

Hakim, Alert B. 2005. *Historical Introduction to Philosophy 5th Edition*. New York: Routledge.

Havelock, Eric. 1976. *Origins of Western Literacy*. Toronto: Ontario Institute for Studies in Education Monograph Series.

Havelock, Eric. 1963. Preface to Plato. Cambridge MA: Harvard University Press.

Hayles, N. Katherine. *The Transformation of Narrative and the Materiality of Hypertext*. 1st ed. Vol. 09. Ohio State UP, 2001. 21–39.

Heylighen, F. and C. Joslyn. 2001. "The law of requisite variety." (http://pespmcl.vub.ac.be/REQVAR.html, accessed March 7, 2019).

Hertzler, Joyce. 1965. A Sociology of Language. New York: Cooper Sq. Pub.

Holy Scriptures. 1951. *The Holy Scriptures According to the Masoretic Text*. Philadelphia: The Jewish Publication Society of America.

Innis, Harold. 1972. *Empire and Communications*. With foreword by Marshall McLuhan. Toronto: Univ. of Toronto Press (Originally pub. by Oxford Univ. Press [1950]).

Iseri, Zeynep Merve and Robert K. Logan. 2016. "Laws of Media, Their Environments and Their Users: The Flip of the Artifact, Its Ground and Its Users." MDPI *Philosophies* 2016, *1*(2), 153–161.

Israel, Nico. 2015. *Spirals: The Whirled Image in Twentieth Century Literature and Art*. New York: Columbia University Press.

Kaplan, Robert. 2000. The Nothing that Is: A Natural History of Zero. Oxford: Oxford University Press.

Kaufmann, Stuart. 1995. *At Home in the Universe*. Oxford UK: Oxford University Press.

Kauffman, Stuart, Robert K. Logan, Robert Este, Randy Goebel, David Hobill and Ilya Shmulevich. 2007. "Propagating organization: an enquiry." Biology and Philosophy 23: 27–45.

Knudson, Eva. 2004. The Circle & the Spiral: A Study of Australian Aboriginal and New Zealand Maori Literature. Amsterdam: Brill Rodopi Publishers.

Kramer, S.N. 1959. *Life Begins at Sumer*. Garden City, N.Y.: Doubleday Anchor Books.

Krems, Jaime Arona, and R.I.M Dunbar. 2013. "Clique Size and Network Characteristics in Hyperlink Cinema: Constraints in Evolved Psychology." *Human Nature* 24.4, 414.

Lamberti, Elena. 2012. Marshall McLuhan's Mosaic. Toronto: University of Toronto Press.

Lettvin, Jerry, Humberto Maturana, Warren McCulloch and Walter Pitts. 1968. "What the Frog's Eye Tells the Frog's Brain". Reprinted from: "The Mind: Biological Approaches to its Functions" Editors: William C. Corning, Martin Balaban, 1968, 233–258.

Lee, N. and Schumann, J.H. 2003. The evolution of language and the symbolosphere as complex adaptive systems. Paper presented at the conference of the American Association for Applied Linguistics, Arlington VA, March 22–25.

Levinas, Emmanuel. 1990. Entre Nous. New York: Columbia University Press.

Levinson, Paul. 1999. *Digital McLuhan: A Guide to the Information Millennium*. New York: Routledge.

Lindia, Matthew S. 2020. Mediums and Imbrication: Toward a Middle-Oriented.

Lindia, Matthew S. and Paul Levinson. 2021. "Tetrads and Chiasmus: A Reclamation of the Tetrad Wheel." *New Explorations: Studies in Culture and Communication 3*.

Logan, Robert K. 1995. *The Fifth Language: Learning a Living in the Computer Age*. Toronto: Stoddart Publishing.

Logan, Robert K. 1997. "The extended mind: understanding language and thought in terms of complexity and chaos theory." Presented at the 7th Annual Conference of The Society for Chaos Theory in Psychology and the Life Sciences at Marquette U., Milwaukee, Wisconsin, Aug. 1, 1997.

Logan, Robert K. 1999. "The Social, Economic and Educational Impacts of Notational Systems." *Semiotica* 125-1/3: 15–20.

Logan, Robert K. 2003. "Science as a Language, the Non-Probativity Theorem and the Complementarity of Complexity and Predictability." In Daniel McArthur & Cory Mulvihil (eds) Humanity and the Cosmos, 63–73.

Logan, Robert K. 2004a. *The Alphabet Effect: A Media Ecology Understanding of the Making of Western Civilization*. Cresskill NJ: Hampton Press (1st edition 1986. New York: Wm. Morrow).

Logan, Robert K. 2004b. *The Sixth Language: Learning a Living in the Internet Age.* Caldwell NJ: Blackburn Press (1st edition 2000. Toronto: Stoddart Publishing; Mandarin Edition ISBN 986-7964-05-5).

Logan, Robert K. 2005. Making Sense of the visual—Is Google the Seventh Language? *Semiotica* 157 (1/4): 345–352.

Logan, Robert K. and John H. Schumann. 2005. The symbolosphere, conceptualization, language and neo-dualism. *Semiotica* 155 (1), pp. 201–14.

Logan, Robert K. 2006. "Neo-dualism and the Bifurcation of the Symbolosphere into the Mediasphere and the Human Mind." *Semiotica* 160; 229–42.

Logan, Robert K. 2007. *The Extended Mind: The Emergence of Language, the Human Mind and Culture.* Toronto: University of Toronto Press.

Logan, Robert K. 2012. What is Information? Propagating Organization in the Biosphere, Symbolosphere, Technosphere and Econosphere. Toronto: Key Publishing.

Logan, Robert K. and Louis Stokes. 2004. *Collaborate to Compete: Driving Profitability in the Knowledge Economy*. Toronto and New York: Wiley (Mandarin Edition ISBN 7-5080-3669-7).

Lord, Albert B. 1960 *The Singer of Tales*. Cambridge, MA: Harvard Univ. Press.

MacArthur, Daniel. 2000. Cosmology, Creation, and Pseudo Questions. Presented at the Humanity and Cosmos Symposium at Brock University.

MacKay, Douglas. 1969. *Information, Mechanism and Meaning*. Cambridge MA: MIT Press.

Maloney, Edward J. 2005. *Footnotes in Fiction: A Rhetorical Approach*. Columbus: Ohio State University Press.

Manovich, Lev. 2001. *The Language of New Media*. Cambridge MA: MIT Press.

Marshack, Alexander. 1964. "Lunar Notations on Upper Paleolithic Remains." *Science* 146, 743–45.

Martin, Frederick. 1995. "Venus and the Dresden Codex Eclipse Table." *Journal for the History of Astronomy, Archaeoastronomy Supplement*, Vol. 26, p. S57.

Martınez-Miranda, Juan and Arantza Aldea. 2005. "Emotions in Human and Artificial Intelligence." *Computers in Human Behavior* 21, 323–341.

Max, D.T. 2009. "The Unfinished." The New Yorker. March 9, 2009.

McLuhan, Eric and Frank Zingrone (eds) 1995. *Essential McLuhan*. Concord Ontario: Anansi.

McLuhan, Marshall. 1951. *The Mechanical Bride, Folklore of Industrial Man.* New York: The Vanguard Press (reissued by Gingko Press, 2002).

McLuhan, Marshall. 1962. *The Gutenberg Galaxy*. Toronto: Univ. of Toronto Press.

McLuhan, Marshall. 1964. *Understanding Media*: Extensions of Man. New York: McGraw Hill.

McLuhan, Marshall. 1967. "Casting my perils before swains". Preface in G. E. Stearn (ed). McLuhan: Hot & Cool. New York: Dial.

McLuhan, Marshall. 1968. "Rhetorical Spirals in Four Quartets". In Dianne Bessai and David Jackel (eds) *Figures in a Ground. Canadian Essays on Modern Literature Collected in Honor of Sheila Watson*. Saskatoon: Western Producer Prairie Books.

McLuhan, Marshall. 1969. Playboy Magazine Interview. Playboy Magazine (March 1969).

McLuhan, Marshall. 1972. "Preface" in Harold Innis's Empire and Communication." Toronto: University of Toronto Press.

McLuhan, Marshall with Nevitt, Barrington. 1972. Take Today: Executive as Dropout. New York: Harcourt Brace Jovanovich.

McLuhan, Marshall, and Quentin Fiore. 1967. *The Medium is the Massage: An Inventory of Effects*. New York: Bantam Books.

McLuhan, Marshall, Quentin Fiore and Michael Agel. 1968. *War and Peace in the Global Village*. New York: Bantam Books.

McLuhan, Marshall. 1970. *Culture is Our Business*. New York: McGraw Hill.

McLuhan, Marshall. 1972a. Foreword to *Empire and Communications* by Harold Innis. Toronto: Univ. of Toronto Press.

McLuhan, Marshall. 1972b. "Canada: The Border Line Case". In David Staines (ed) The Canadian Imagination: Dimensions of Literary Culture. Cambridge MA: Harvard University Press, 226–48.

McLuhan, Marshall and Eric McLuhan. 2011. In Marshall McLuhan and Eric McLuhan (eds) Theories of Communication.

McLuhan, Marshall, and Eric McLuhan. 1988. Laws of Media: The New Science. Toronto: University of Toronto Press.

McLuhan, Marshall. 1975. "Communication: McLuhan's Laws of Media." *Technology and Culture* 16 (1), 74–78.

McLuhan, Marshall. 1977. "Laws of Media." English Journal 67 (8): 92–94. Also published *Et Cetera* 34 (2): 173–179.

McLuhan, Marshall. 1987. "To Johnathan Miller." In M. Molinaro, C. McLuhan & W. Toye (eds) *Letters of Marshall McLuhan*. Toronto: Oxford University Press.

McLuhan, Marshall. 1999. *The Medium and the Light: Reflections on Religion*. Edited by Eric McLuhan & Jacek Szklarek. Toronto: Stoddart.

McLuhan, Marshall. 2011. *Counterblast*. Berkeley CA & Hamburg: Ginko Press (Originally published in 1954).

McLuhan, Marshall and Robert K. Logan. 1977. "Alphabet, Mother of Invention." Etcetera 34: 373–83.

McLuhan, Marshall, and Eric McLuhan. 1988. *Laws of Media: The New Science*. Toronto: University of Toronto Press.

McLuhan, Marshall; McLuhan, Stephanie; Staines, David. 2003. *Understanding Me: Lectures and Interviews*. Toronto: McClelland and Stewart.

McLuhan, Marshall and Bruce R. Powers. 1989. *The Global Village: Transformations in World Life and Media in the 21st Century*. New York: Oxford University Press.

McLuhan, Marshall and Wilfred Watson. 2011. From Cliché to Archetype. Berkeley CA: Ginko Press (originally published in 1970 by Viking Press).

Miller, Lynn C. 1993. "Writing is hearing and saying: Gertrude Stein on language and in performance." *Text and Performance Quarterly* 13(2), 154–167.

Molinaro, Matie, Corrine McLuhan, and William Toye (eds). 1987. *Letters of Marshall McLuhan*. Toronto: Oxford University Press.

Mundaca, Marie. 2009. "The Influence of Anxiety: Wading In", *Hipster Bookclub*. June 2009.

Nabokov, Vladimir. 1962. Pale Fire: A Novel. New York: G.P. Putnam's Sons.

Nelson, Ted. 1965. "Complex Information Processing: A File Structure for the Complex, the Changing and the indeterminate." In Jill Walker Rettberg (ed). Association for Computing Machinery (ACM) '65 Proceedings of the 1965 20th National Conference (http://elmcip.net/node/7367).

Nelson, Theodor. 1974. *Computer Lib: You Can and Must Understand Computers Now; Dream Machines: New Freedoms Through Computer Screens—A Minority Report*. Self-published. ISBN 0-89347-002-3; Tempus Books/Microsoft (2nd ed 1987). ISBN 0-914845-49-7.

Nelson, Theodor. 1982. *Literary Machines*. Sausolito CA: Mindful Press.

Neugebauer, Otto. 1969. *The Exact Sciences in Antiquity, 2 ed.* Mineola NY: Dover Publications.

Nonaka, I., and H. Takeuchi. 1995. *The Knowledge Creation Company*. New York: Oxford University Press.

Ong, Walter. 1991. Orality and Literacy. London: Routledge.

Ong, Walter. 1996. *Interview.* In Michael Kleine and Fredric G. Gale (eds) "The Elusive Presence of the Word: An Interview with Walter Ong." Composition FORUM (1996), 65–85.

Pais, Abraham. 1989. "George Uhlenbeck and the Discovery of Electron Spin." *Physics Today* 42, 12, 34.

Parry, Milman. 1993. *The Making of Homeric Verse: The Collected Papers of Milman Parry.* Adam Parry (ed). Oxford UK: Oxford University Press.

Paivio, A., and I. Begg. 1981. Psychology of Language. Englewood Cliffs, N.J.: Prentice-Hall.

Peirce, Charles Sanders. 1932. In Charles Hartshorne and Paul Weiss (eds) *Collected Papers of Charles Sanders Peirce, Volumes I and II: Principles of Philosophy and Elements of Logic*. Cambridge MA: Harvard University Press.

Petrie, W. F. Flinders. 1906. *Researches in Sinai*. London: John Murray.

Pietropaolo, Domenico. (1995) "Vichian Ascendancy in the Thought of Marshall McLuhan". *New Vico Studies* 13: 55–62.

Pinker, Steven. 2003. "Language as an adaptation to the cognitive niche". In M. H. Christiansen and S. Kirby (eds) *Language Evolution: The States of the Art*. Oxford: Oxford University Press.

Popper, Karl. 1959. *The Logic of Scientific Discovery*. London: Hutchinson & Co.

Potin, Jason. 2014. "The neuroscientist Antonio Damasio explains how minds emerge from emotions and feelings." Technology Review June 17, 2014, (https://www.technologyreview.com/s/528151/the-importance-of-feelings, accessed June 27, 2019).

Pruska-Oldenhof, Izabella. 2011. "'Man as the Medium': McLuhan, Etrog, and Spiral Archetype." *Ultimate Reality and Meaning (UTP)* 34 (3–4), 257–66.

Purce, Jill. 1971. "The Spiral in Art: Aspirations Towards Eternity." *The Structurist* 11.

Purce, Jill. 1980. *The Mystic Spiral: Journey of the Soul (Art and Imagination)*. London: Thames & Hudson (Reissue edition).

Quart, Alissa. 2005. "NETWORKED: Dysfunctional families, reproductive acts, and multitasking minds make for Happy Endings" *Film Comment* 41.4, 48–51.

Rushkoff, Douglas. 2016. *Throwing Rocks at the Google Bus*. New York: Portfolio/Penguin.

Sabelli, Hector and Louis H. Kauffman. 2013. "The Biotic Logic of Quantum Processes and Quantum Computation." In Franco Orsucci and Nicolette Sala (eds) Complexity Science, Living Systems, and Reflexing Interfaces: New Models and Perspectives. Hershey PA: IGI Global. Chapter 7, 112–183.

de Saussure, F. 1967. Grundfragen der allgemeinen Sprachwissenschaft. [Basic Questions of Linguistics] 2d ed. Berlin.

Sapir, E. 1921. Language: An Introduction to the Study of Speech. New York: Harcourt, Brace & World Inc.

Schmandt-Besserat, Denise. 1978. "The Earliest Precursor of Writing". *Scientific American* 238.

Schmandt-Besserat, Denise. 1981. From Tokens to Tablets: A Re-Evaluation of the So-Called Numerical Tablets. *Visible Language* 15.

Schmandt-Besserat, Denise. 1984. Before Numerals. Visible Language 18.

Schmandt-Besserat, Denise. 1992. *Before Writing: Vol. 1. From Counting to Cuneiform*. Houston: Univ. of Texas Press.

Schneider, Thomas. 1990. "The Relative Chronology of the Middle Kingdom and the Hyksos Period (Dyns. 12–17)" In Erik Hornung, Rolf Krauss & David Warburton (eds) Ancient Egyptian Chronology *Handbook of Oriental Studies*. Leiden: Brill: 2006.

Schumann, John H. 2003a. "The evolution of language: What evolved?" Paper presented at the Colloquium on Derek Bickerton's Contributions to Creolistics and Related Fields, The Society for Pidgin and Creole Linguistics Summer Conference, Aug. 14–17, University of Hawaii, Honolulu.

Schumann, John H. 2003b. "The evolution of the symbolosphere". *Great Ideas in the Social Sciences Lecture*, UCLA Center for Governance, Nov. 21.

Shiryayev, A.N. (Editor). 1993. Selected Works of A.N. Kolmogorov: Volume III: Information Theory and the Theory of Algorithms (Mathematics and its Applications). New York: Kluwer Academic Publishing.

Sitney, P. Adams. 2002. *Visionary Film: The American Avant-Garde 1943-2000*. 3rd Edition. Oxford: Oxford University Press.

Smith, Frank. 1982. Writing and the Writer. New York: Holt, Rinehart & Winston.

Smithson, Robert. 1996. "The Spiral Jetty (1972)" in Jack Flam (ed) *Robert Smithson: The Collected Writings*. Berkeley and Los Angeles CA: University of California Press.

Steinsaltz, Adin. 1976. The Essential Talmud. New York: Basic Book.

Stein, Gertrude. 1990. Selected Writings of Gertrude Stein. New York: Vintage Press.

Stein, Gertrude, Clement Hurd & Thacer Hurd. 2013. New York: Harper Design.

Stubbs, Michael. 1980. Language and Literacy. London: Routledge & K. Paul.

Tomasello, Michael. 1999. *The Cultural Origins of Human Cognition*. Cambridge, MA: Harvard University Press.

Thieberger, Nick. 2012. The Oxford Handbook of Linguistic Fieldwork. Oxford UK: Oxford University Press.

Usher, Abbot Payson. 1959. *A History of Mechanical Inventions*. Boston: Beacon.

Vygotsky, Lev S. 1962. *Thought and Language*. Cambridge, MA: M.I.T. Press.

Vidal, Gore. 1978. *Kalki*. New York: Random House.

Wallace, David Foster.1996. *Infinite Jest: A Novel*. Boston: Little, Brown.

Wallace, David Foster. 2005. *Consider the Lobster*. Boston: Little, Brown.

Wallace, David Foster. 2009. *This is Water: Some Thoughts, Delivered on a Significant Occasion about Living a Compassionate Life*. New York: Little, Brown.

Watson, J.D. and F.H.C. Crick. 1953. "A Structure for Deoxyribose Nucleic Acid." *Nature* 171, 737–738.

Wolf, Gary. 2015. "The Curse of Xanadu" *Wired*. 1 June 1995 (Can be accessed https://www.wired.com/1995/06/xanadu/).

Zerubavel, Eviatar. 1989. *The Seven Day Circle: The History and Meaning of the Week*. Chicago: University of Chicago Press.

Zeynep Merve Iseri and Robert K. Logan. 2016. "Laws of Media, Their Environments and Their Users: The Flip of the Artifact, Its Ground and Its Users." MDPI Philosophies 2016, 1(2), 153-161; doi.10.3390/philosophies1020153

The manufacturer's authorised representative in the EU is Springer Nature Customer Service Centre GmbH, Europaplatz 3, 69115 Heidelberg, Germany. If you have any concerns regarding our products, please contact ProductSafety@springernature.com

Printed and bound by CPI Group (UK) Ltd, Croydon, CR0 4YY
30/09/2026
02225325-0008